The Biometric Industry Report

Forecasts and Analysis to 2006

Second edition

ELSEVIER

UK	Elsevier Science Ltd, The Boulevard, Langford Lane, Kidlington, Oxford OX5 1GB, UK
USA	Elsevier Science Inc, 360 Park Avenue South, New York, NY 10010-1710, USA
JAPAN	Elsevier Science Japan, Tsunashima Building Annex, 3-20-12 Yushima, Bunkyo-ku, Tokyo 113, Japan

Author: Mark Lockie, Ideal Media Consulting

Programme Editor: Roisin Reidy

First Edition published January 2001
Second Edition published November 2002

British Library Cataloguing in Publication Data
A CIP Catalogue record for this book is available from the British Library.

ISBN: 1 85617 394 1

Published by
Elsevier Advanced Technology
The Boulevard, Langford Lane, Kidlington, Oxford OX5 1GB, UK
Tel: +44 1865 843000
Fax: +44 1865 843971

Typeset by Land & Unwin (Data Sciences) Ltd, Bugbrooke
Printed and bound in Great Britain by Biddles Ltd, Guildford and King's Lynn

Contents

List of Tables

List of Figures

Acknowledgements

The author would like to express thanks to the many individuals who helped in the preparation of this report. In particular thanks go to Wendy Atkins at Ubiquitous Media, Stuart Lockie at Cielok Enterprises, Dr Brigitte Wirtz at Infineon Technologies and Martin George at ID Craft.

1 Introduction

The second edition of The Biometric Industry Report is published at a turbulent period in the biometric industry's history. Difficult world events have conspired to sway the industry from buoyant growth to recession and now back to sharp growth. The end result has been a delay in predicted industry development from the last report by a period of just under two years.

Still there is much to be optimistic about as the industry looks forward. Excluding any world changing events, which cannot be predicted, the industry is forecast to break through the US$1 billion barrier within the timeframe of this report (2000–2006) reaching US$1.87 billion by 2006. In 2002 the industry is forecast to make just over US$260 million.

The main driving forces behind this strong growth are expected to be the deals coming from the government and travel/immigration vertical markets, primarily to enhance homeland security. Physical access control applications will continue their steady march forward, but strong growth is expected to return to the computing sector, where biometrics will increasingly be used to help secure PCs, networks and intranets.

From a technological standpoint, the biometric industry is dominated by the fingerprint sector, which controls more than 40% of the overall market. This position is highly unlikely to change as the technology benefits from a good track record, cheap devices, high accuracy and small size. However, there has been strong growth by facial and iris recognition technologies, as well as the middleware sector since the last report was compiled in 2000.

Primarily, facial recognition has surged forward since September 11th, 2001, as a potential technology to help in the fight against terror, particularly when used in surveillance applications. Iris recognition is now established as the highest accuracy device on the market today, and although it is still a pricey solution, it provides high security with virtually zero false matches – and this could prove critical for some of the large-scale implementations now being planned.

Geographically, North America has reasserted itself as the primary hub of the biometric industry. Driven forward by the terrorist attacks last year, the region has experienced a flurry of biometric activity, with the promise of much larger implementations to come. Asia and Europe had been North America's closest rivals, but they will continue to play second fiddle for the foreseeable future. Meanwhile Asia will overtake Europe to become the world's second biggest region for biometrics.

The industry continues to undergo radical structural change. Throughout 2002 there was news of biometric companies faltering on economic grounds. Inevitably, this trend is expected to continue until the industry comprises a much smaller number of elite companies with the resources and expertise to meet the demands of large implementations that will be the lifeblood of the industry in the future. There are still some investors, albeit not as many as in days of old, that are keen to put money into the biometric companies they believe could make the big time – as such the shakeout of companies will not happen overnight, but over the next 24 months.

Notably, there are players in the industry that are beginning to focus on specific niche markets, and this is a trend for the better. There are also going to be an increasing number of resellers, providing outlets for the elite companies' products. This trend is evident today and overall it actually swells the ranks of the industry to above 300 companies.

Developments on the standards front have been prolific over the last two years, as the BioAPI standard became widely accepted and was fast-tracked for ISO approval. Ongoing work now seems to be aimed at accelerating the integration of existing standards into products and solutions, rather than developing new standards.

This report provides the reader with a thorough and easy-to-understand snapshot of the industry in the third quarter of 2002, and uses well defined forecasting techniques to assess the future direction of the industry up until 2006. It should help readers to assess the likely development of the industry as a whole, as well as providing detailed information on the specific issues within each technology and market covered. The report also contains in-depth company profiles, which are our pick of the leading and strategically interesting companies in the industry today. There is also a comprehensive address book broken down by different categories and appendices that include a complete and updated introduction to biometrics.

1.1 Summary of Contents

- Chapter 2 looks at the industry as it stands today. This includes an overview of revenue, structure, consolidation and the issues facing the industry, such as privacy, standards, prices and spoofing.
- Chapter 3 is the heart of the report, forecasting industry development to 2006. This includes a comprehensive overview of each main technology from a market and technical standpoint, as well as in-depth analysis on each main vertical market. For each of these there will also be a discussion of all the key drivers and detractors affecting revenue growth over the next few years. Also forecast is the development of the industry's horizontal and geographic markets.

- Chapter 4 is dedicated to in-depth company profiles of some of the industry's leading players, as well as some players that are interesting from a strategic point of view. For each company we look at its history, technology, product range, strategy and financial information, where available.
- Chapter 5 is a fully up-to-date address book containing well over 300 companies. As well as an A–Z listing it also looks at the industry by geographic region, technology and country.
- Chapter 6 comprises the appendices, including a complete beginners' guide to biometrics, an extensive glossary of terms and an in-depth overview of standards.

2 The Biometric Market in 2002

2.1 Introduction

The biometric market finds itself at a critical point of its development in 2002. Following many years of promise, the time has now come for the technology, and the vendors who supply it, to deliver. This is no exaggeration. Terrorist events in the USA in September 2001 focused the attention of the world's governments on the technology, as they seek ways to improve intelligence and strengthen their borders. If biometric vendors can deliver effective solutions over the next few years, then the industry's future is assured and growth will move strongly upwards. If the systems turn out to be flawed and ineffective then damage to the industry's credibility may be unrecoverable.

Much work has been done in the last few years and this report, while highlighting the difficulties ahead, is positive in its assessment of the industry's chances going forward. After all no other technology has the ability to say with authority that a person is really the person they claim to be. It is also one of the only technologies that can be used to automate surveillance systems or to discover people fraudulently using multiple identities.

This chapter will provide an insight into the state of today's industry. It will begin with an overview of the methodology used to create the report's statistics. Following this will be an assessment of the industry's structure, consolidation trends, price developments and a look at some of the issues such as privacy and standards.

2.2 Methodology

Market figures in this report were produced using a methodology that was carefully designed to create an accurate picture of the biometric industry in 2002 and for the timeframe forecast in this report (to 2006).

The first step was to assess what the industry wanted from such a report. It became clear that a breakdown of the industry by technology, horizontal market (what function the biometric performs), vertical market (the sector where the biometric is being used) and geography was needed. In addition, the in-depth company profiles and address book were found to be extremely useful in the first edition of this report, so they were repeated, but are now more comprehensive and useable.

Approximately 75 major companies producing biometric products were selected, as these companies generate the vast majority of revenue within the industry. Although there are more companies in the industry than this, the ones omitted are still believed to be establishing themselves or acting as resellers of other companies' solutions.

Each of these companies received an in-depth questionnaire that was designed to generate all the required data sets for this report's statistics. The figures received from these companies contributed to the calculation of the industry size between 2000–2002. Responses were received from 70% of companies targeted, and it was possible to estimate the remainder through market knowledge, published literature and other indicators.

In order to forecast up to 2006, a number of factors were considered. First, a large number of interviews were performed to research the drivers and detractors in each vertical market. An extensive search was also undertaken to discover all relevant market information about each of these verticals. Using this information, as well as the historical data from the questionnaires it was possible to forecast the likely progression of each market over the next four years.

Following on from the vertical market forecasts, each technology was extensively researched and again driver and detractor information was assessed to see how the technology would likely develop. By looking at the vertical markets and estimating the likely use of each technology for that vertical market (based on previous installation history and likely future technological developments) it was possible to build a picture of the technology landscape in the future. The horizontal and geographical market information was assessed in a similar way.

2.3 Today's Market

The biometric industry has been on a revenue roller coaster ride over the last three years. Impressive growth in 1999 had set the industry up for a bumper year in 2000. Then the unthinkable happened. A dot com crash shattered stock market confidence. Soon afterwards an economic slowdown in the USA, and subsequently in a large portion of the world, took hold.

Going into 2000, the biggest driver for growth in the biometric industry had been IT security, but this was one of the areas hardest hit in the uncertain economic climate. A telecoms industry slump, although not so significant for biometrics, also did not help matters.

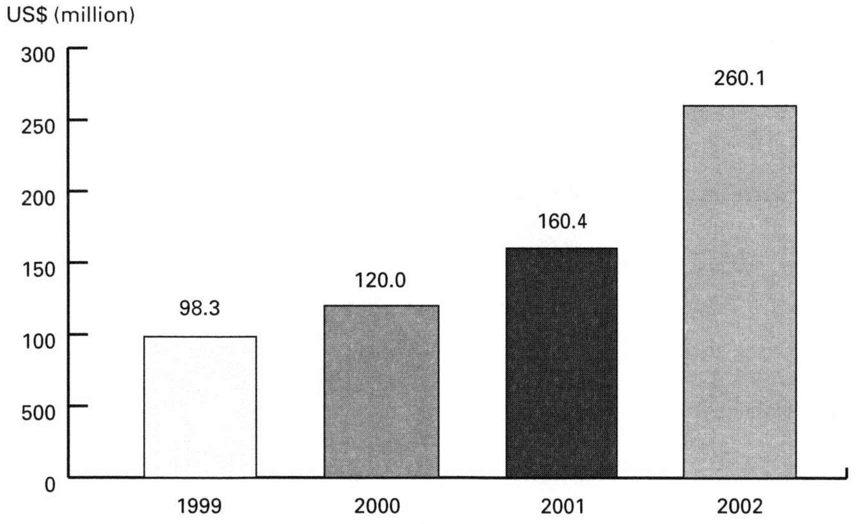

Figure 2.1 Total Industry Revenue 1999–2002 (US$ million)

The first edition of this report was compiled in the first half of 2000 and bullish forecasts had seemed entirely reasonable. But as the extent of the slowdown became clear, many commercial companies battened down the hatches and put the brakes on spending, especially in areas such as security, and particularly if they were IT related.

The economic landscape hardly improved in 2001 and it seemed almost certain that biometrics would have to sit out the economic downturn before returning to better times. Then September 11th occurred. Terrible events as they were, they provided a significant boost to the industry both in terms of revenue and in terms of raised awareness.

Today, almost every supplier will tell you that they will be securing impressive contracts in the new security conscious world. Clearly not everyone can be a winner, but meanwhile, the industry has returned to a more sustainable growth rate of around 60% per annum.

Figure 2.2 shows the industry's market share statistics for 2002. Clearly fingerprint technology is the leading technology, while the USA remains the dominant regional player. Meanwhile, physical access control applications have the edge over IT security installations and the leading vertical market is the general commercial industry followed by the government sector.

2.3.1 Structure

The biometric industry is a diverse entity. With six main technologies, five main horizontal markets, nine vertical markets and five major geographical markets, there are many variations to consider when writing an industry 'state of play' report.

The biometric industry is still emerging and even in 2002 a moderately sized deal can affect overall industry performance from month to month. There are some companies that are now beginning to find their place in the market, but a

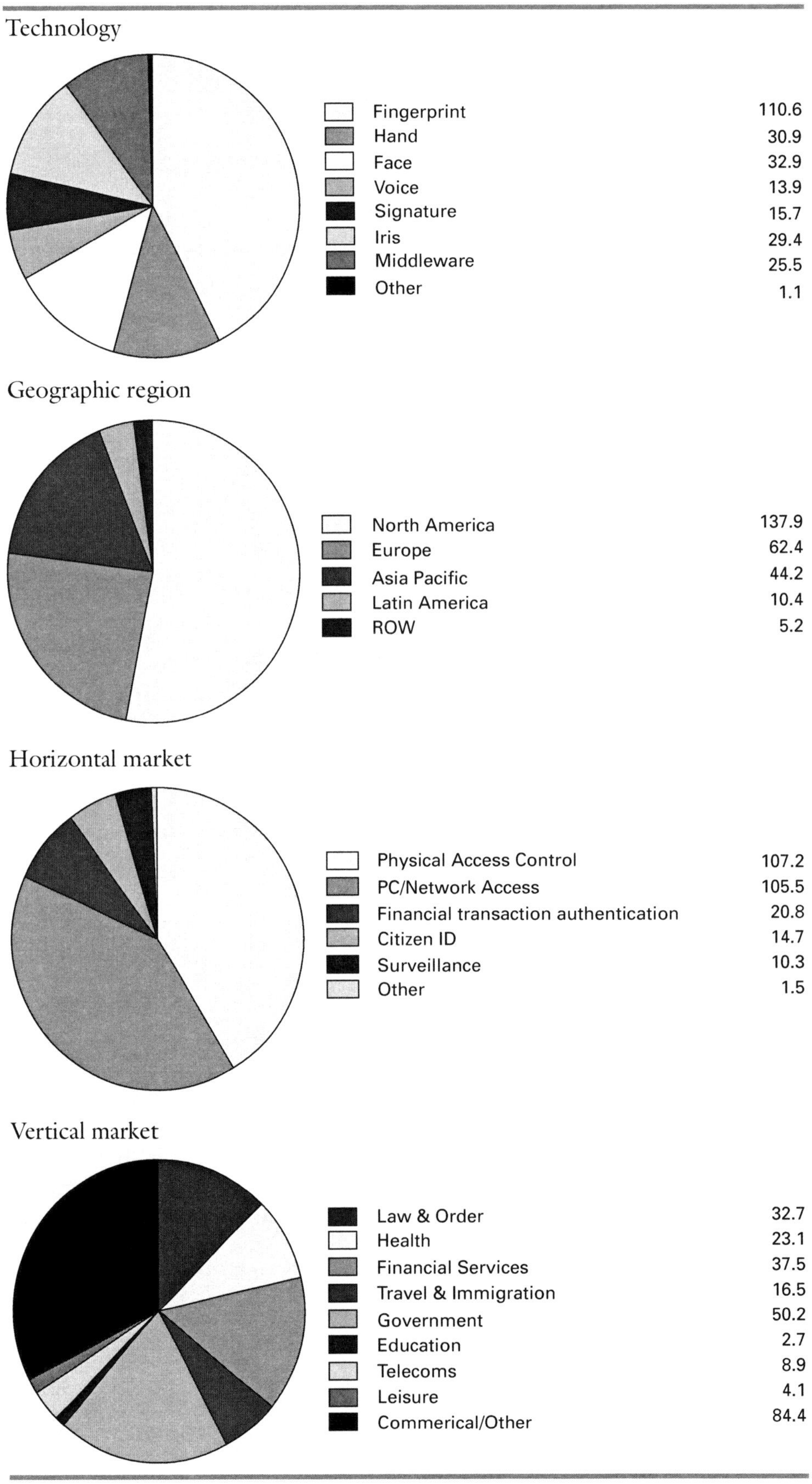

Figure 2.2 Market Share of Each Technology, Geographic Region, Horizontal Market and Vertical Market in 2002 (US$ million)

large number of others are still struggling to define a *raison d'être* that will guarantee their future success in the market. Severe financial concerns are now also starting to affect most players in the industry.

There are more than 300 companies working in the biometric space – as evidenced by the industry directory included in this report. However, the number of companies providing the sensors, matching algorithms, middleware and devices is much smaller at between 100–150. Many of the other companies are resellers, original equipment manufacturers (OEMs) and integrators – not to mention media groups, industry associations and academic establishments.

In fact the number of value added resellers (VARs) is proliferating in the industry. This is an important trend as the ever-smaller number of companies producing the base products cannot hope to reach the markets and geographical locations that can be served by an efficient network of competent VARs.

Of the different groups in the industry, some produce the sensors, the software and the devices as an overall package, while other companies are expert in just one of these segments. Some of the companies are also targeting specific niche markets with their products, while others still advertise their wares for almost any vertical market thinkable.

As part of this research report a number of statistics were gathered to enable the calculation of the industry's 'average' company (see Table 2.1). The most striking statistic is the drop in staff numbers. This is most likely explained by the number of redundancies being made by companies in the market at present in a bid to cut costs.

Table 2.1 The Average Biometric Company

	2002	2000
Average company size	33 employees	46 employees
Average company age	8 years	7 years
Average revenue US$	4.8 million	3.2 million
Average number of R&D staff	13	–
Average percent of total revenue spent on R&D	39%	59%

2.3.2 Consolidation

Almost every week it is reported that a biometric company is shutting its doors. This is a sad trend that looks set to continue for the foreseeable future.

In the tough economic climate at present, many of the industry's suppliers are struggling to keep their heads above water. Unfortunately, the lifeblood of the industry in previous years was the financial investor. Today, investors are an extremely rare commodity, so this critical lifeline for vendors has been severed.

In the absence of this income source, many of the industry's players are facing the possibility of closure. At current cash burn rates, many have less than one year's worth of money remaining. So other than going into liquidation, the only options left are to make cost cuts (including staff redundancies) or to refocus the business. In the past, going public was also an option to raise more cash, but in light of the technology market crash, this option seems most unfeasible for the time being.

For some companies income levels will rise enough in the next 12–24 months to prevent their closure, but the line between success and failure is a fine one at present and strong management skills will be required to steer companies away from the precipice.

It is likely that the biometric landscape by 2006 will be a group of 20–30 elite biometric companies each with numerous resellers and integrators of their products. While this development is a sad trend, it will enable the industry to emerge leaner and more robust than ever and this is essential if the world at large is to trust the industry with its investment dollars again – and will be more reassuring for potential customers.

2.3.3 Other Market Issues

Away from the economic trials and tribulations of the biometric industry, there are numerous issues that must still be resolved by the vendors. The most predominant of these are developing standards, privacy issues and guarding against spoofing.

In the past, the price of biometric systems was a significant issue, but over the last two years this has receded somewhat as the price of many of the systems on the market have dropped to acceptable levels. This does depend on the market however. In access control applications, for example, biometrics, such as iris recognition, are still fairly expensive compared with the non-biometric competition.

Standards

Standards work has moved forward in leaps and bounds over the last two years. A full description of the development work is given in section 6.3.

Most importantly, the BioAPI standard has now received backing from a large proportion of the biometric industry's players and has been accepted as an official ANSI standard. More exciting are the latest moves to fast track the standard towards becoming an ISO standard. Development of the main standards governing biometric data interoperability, such as the BioAPI BIR and CBEFF, has also moved forward.

The imminent launch of Microsoft's new operating system with a Biometric API (not the BioAPI) had been expected to cause a significant rift in the biometric industry. But now this is not believed to be a significant problem. This is because it is likely that an appropriate translation layer between BioAPI and the Microsoft BAPI will be established.

Privacy

Privacy issues have always been a thorn in the biometric industry's side. This is despite the fact that biometrics can be effectively used to guard a person's privacy, particularly when a person carries their biometric template on a smart card. For many applications this smart card arrangement is not available, however, and privacy advocates (as well as a portion of the general public) still feel uncertain as to the technology's use.

In particular the use of biometrics in surveillance applications has created massive problems over the last year with much bad publicity. Further details on the installations and privacy issues surrounding facial recognition systems are given in Section 3.3.3.

A new legislative Bill was recently proposed in the USA designed to set guidelines for the possession and handling of biometric data (the Biometric Identifier Privacy Act (A-2448)). It still must pass through the standard legislative procedures but if enacted it would prevent a person from selling, leasing, transferring or disclosing biometric data unless the following conditions are met:

- the individual consents to the sale, lease or disclosure;
- the sale, lease or disclosure completes a financial transaction requested or authorized by an individual;
- the sale, lease or disclosure is required or permitted by federal or state law;
- the sale, lease or disclosure is made by a law enforcement agency for a law enforcement purpose.

Any violation would be liable for a penalty of up to US$25,000 for each violation.

On a lighter note, two polls looking at public opinion surrounding privacy issues recently indicated that the majority of US citizens were happy to use biometrics in their daily lives, with more than 70% in favour of carrying identification cards with fingerprints. Only 11% of participants in one poll perceived biometrics as an invasion of privacy.

Spoofing

This issue is comprehensively covered throughout the next chapter as part of the discussions of each technology. However, this is a serious issue that has reared up in the last 18 months, as increasing numbers of researchers have attacked biometric technologies using relatively simple methods.

It appears that no biometric technology is immune from this problem – although the behavioural biometrics, such as dynamic signature verification and voice verification, have not yet been subject to any concerted attack.

The whole concept of biometrics is to enhance security, so the idea that simple attack methods can be used to bypass this security makes for alarming reading. The industry has reacted to the claims and hopefully through additional research and development this issue can be dealt with quickly. Inevitably, however, this will now become a continual battle between hacker and vendor.

3 Biometric Market Forecasts

3.1 Introduction

Revenue growth in the biometric industry experienced a slowdown in the two years since the first edition of this report was published in 2000. Despite the fact that the industry was headed strongly upward in 1999, economic slowdown, the dot com crash and the telecoms collapse served to severely dampen industry growth in the second half of 2000 and throughout 2001.

There has been a pick up over the last year, mainly because of the after effects of September 11th, 2001 and the extraordinary interest in security that followed. It could be said that while all economic factors were dampening demand in the biometric industry, the events of September 11th restored a degree of balance to the market. Nevertheless the industry has been set back and isn't expected to break the magic US$1 billion mark until 2005.

This chapter will provide forecast figures for each major technology, horizontal market, vertical market and geographic market in the biometric industry.

The horizontal market focus is a new addition to the report. It essentially looks at the biometric industry in terms of what function any particular biometric technology is performing. This includes physical access control, surveillance, PC access, transaction security and so on.

The vertical markets in this report are simply the sector that a biometric is being used in, such as the healthcare, government or commercial sector.

By way of a further explanation consider the following – a casino (vertical market – leisure) based in South Korea (geographical market – Asia) uses a biometric access control system (horizontal market – physical access control) based on fingerprint recognition technology (technology market – fingerprint).

In the following pages, each technology and vertical market will be considered in greater depth than revenue trends alone. They will also be looked at in terms of market position, technological development, and what factors are driving or detracting from their future.

3.2 Total Industry Revenue

The biometric industry has experienced some very tough times over the last three years. As highlighted above, economic recession and a collapse in the IT sector put many potential biometric projects neatly back on the shelf as companies cut back on all luxury projects.

Table 3.1 tells the story quite effectively in numbers. Growth still occurred during the lean years of 2000 and 2001 (after all the industry was starting from such a low base that even a handful of large deals would help provide overall industry growth), but it was severely curtailed.

Table 3.1 Industry Growth Rate 1998–2006

Year	Revenue (US$ million)	Growth rate
1998*	52.9	
1999*	98.3	86%
2000	120.0	22%
2001	160.4	34%
2002	260.1	62%
2003	423.6	63%
2004	707.3	67%
2005	1131.6	60%
2006	1867.2	65%

**Source:* The Biometric Industry Report – Market and Technology Forecasts to 2003, 1st Edition

Only following September 11th did the industry begin to pick up in terms of increased interest from outsiders and subsequently revenue. Many companies re-established their shelved projects and the dollars have begun to flow.

Although there are many people saying the contracts arising because of September 11th are slow in coming, annual growth has now returned to more acceptable (and sustainable) levels at around 60%. It is true that the really big government contracts are taking a long time to appear, but elsewhere (partly because of the increased awareness of biometrics), the growth has been notable.

The figures in this report were calculated in the middle of 2002, so there is always uncertainty surrounding the second half and beyond, but the green shoots of recovery are evident.

Barring any disasters of the magnitude of September 11th, the growth of the industry to 2006 will be strong and steady. While the IT security market still needs time to recover, much of the new money is coming from government interest, the aviation sector and the law and order markets.

The total industry revenue statistics above are the industry's headline figure. Forecasting revenue development into the future is not calculated by looking at previous revenue levels – although these are clearly important. Rather it is

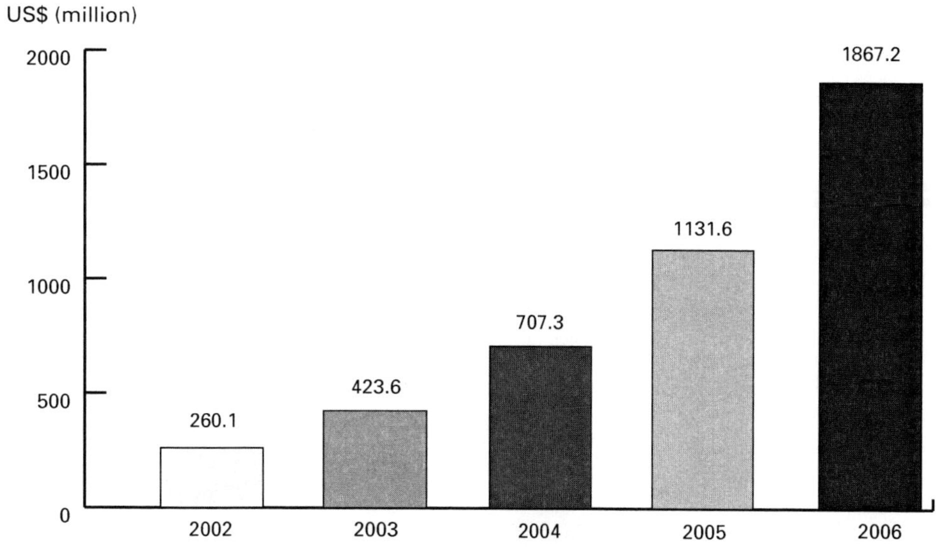

Figure 3.1 Total Industry Revenue Forecast 2002–2006 (US$ million)

calculated through an in-depth assessment of individual vertical markets and their drivers and detractors, as well as by looking at other factors, such as the general economic outlook, the overall demand for security products and the price fluctuation of biometric products.

3.3 Technology Market Assessment and Forecasts

Biometric technologies are undergoing a period of rapid development, as interest from the outside world focuses in on the advantages *and disadvantages* that the technology brings in terms of security, economics and convenience.

Each of the technologies has their own particular combinations of features and drawbacks that shape their performance, acceptability, cost, and ultimately their place in the market.

By assessing all the major horizontal and vertical markets in the industry and understanding the technical requirements of each, it is possible to get a fairly accurate picture of how the different technologies will fare.

Certain technologies will inevitably do well in certain markets. For example facial recognition will do well in surveillance applications and fingerprint recognition will be the leader in the PC market.

From Figure 3.2 it can be clearly seen that the fingerprint market is set to extend its dominance of the market. However, facial recognition is also likely to do well, closely followed by middleware, iris and dynamic signature verification.

Middleware is not exactly a biometric technology. It is authentication management software that generally can accept input from a number of different biometric

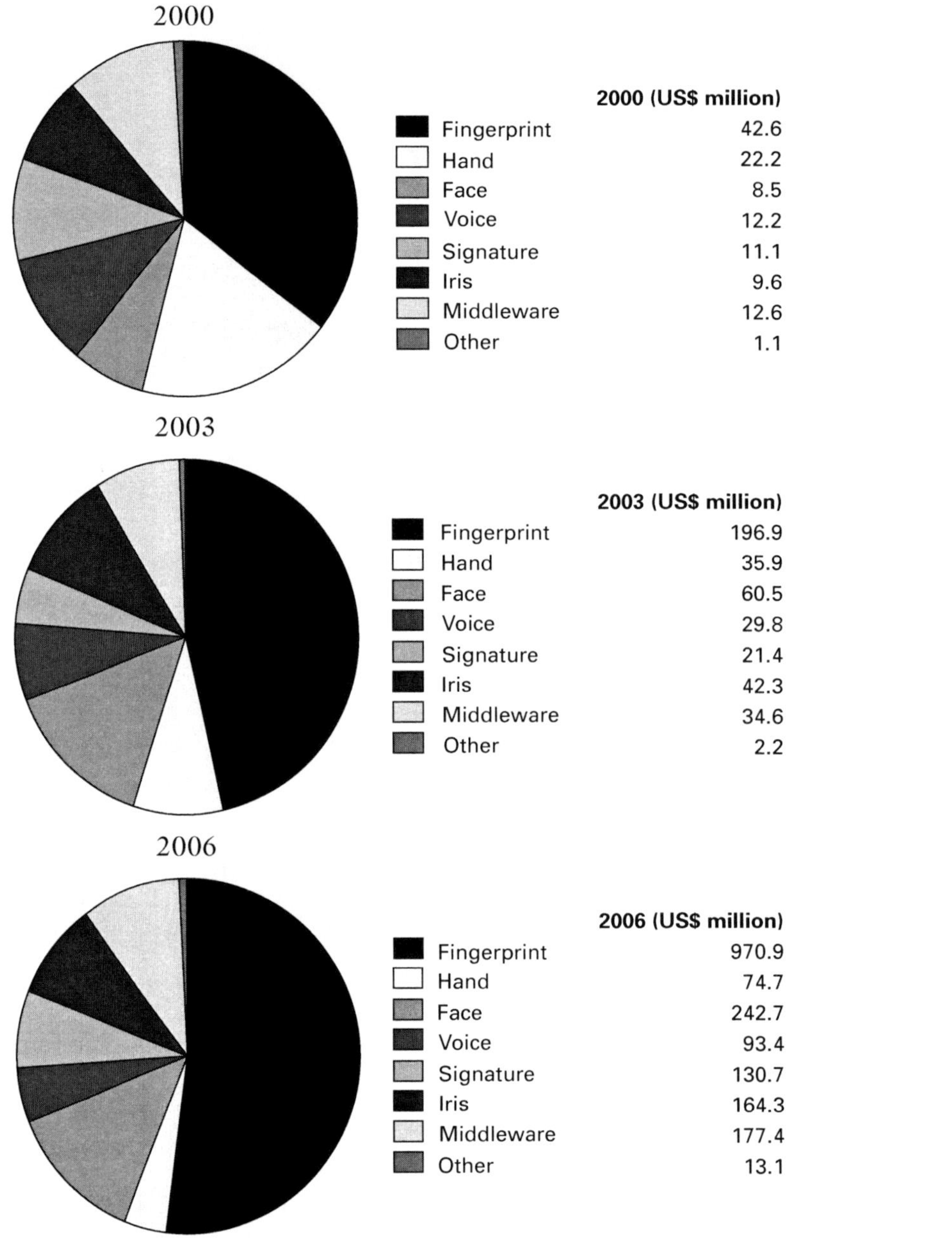

Figure 3.2 Market Share of Technologies in 2000, 2003 and 2006 (US$ million)

modalities. Therefore, although it is included as an important growing segment in this report's analysis, it is not covered in depth as a specific technology.

As well as an estimate of revenue generation for each of the major technologies, the following sections will highlight technical and market considerations, as well as giving in-depth driver/detractor information.

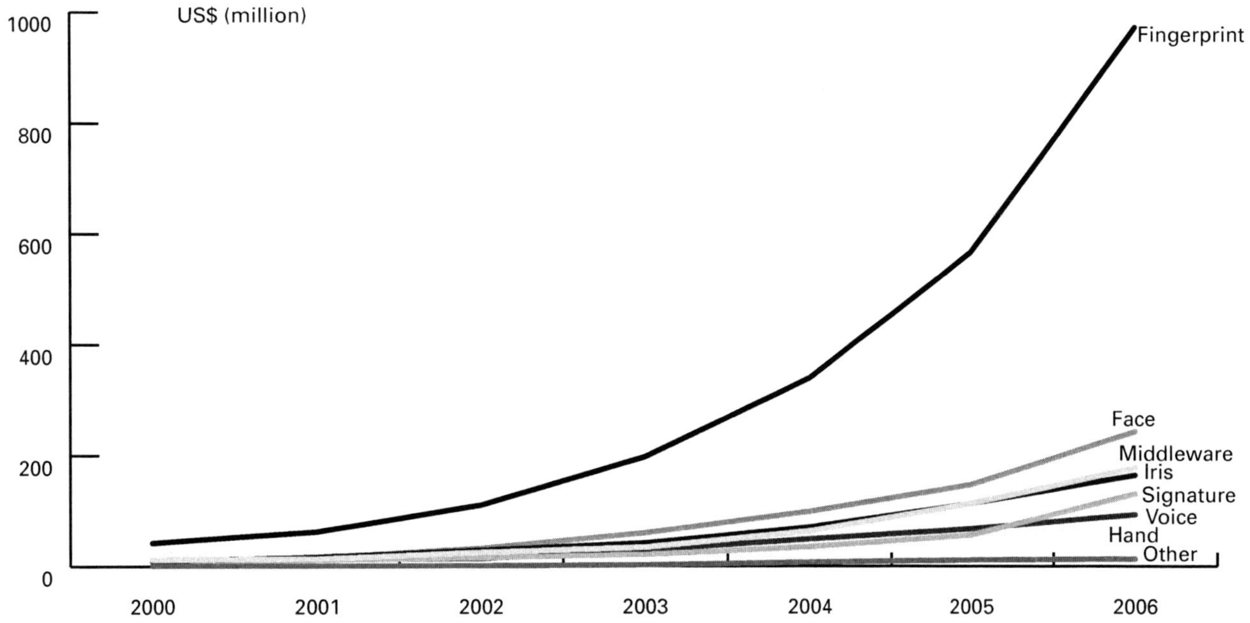

	2000	2001	2002	2003	2004	2005	2006
Fingerprint	42.6	63.0	110.6	196.9	339.5	565.8	970.9
Hand	22.2	25.4	30.9	35.9	42.4	56.6	74.7
Face	8.5	16.0	32.9	60.5	99.0	147.1	242.7
Voice	12.2	12.6	13.9	29.8	49.5	67.9	93.4
Signature	11.1	11.3	15.7	21.4	35.4	56.6	130.7
Iris	9.6	17.4	29.4	42.3	70.7	113.2	164.3
Middleware	12.6	13.9	25.5	34.6	63.7	113.3	177.4
Other	1.1	0.8	1.1	2.2	7.1	11.2	13.1

Figure 3.3 Revenue Forecast by Technology 2000–2006 (US$ million)

3.3.1 Fingerprint Recognition Market

Fingerprint technology is the most widespread biometric in the world and this position is likely to be extended in the future. There are very few markets where the fingerprint sensor isn't a strong contender as it combines accuracy and convenience with a low price tag and small size, not to mention a wealth of installations to prove the technology's viability.

Numerous fingerprint-centric companies exist in the market today (over 80) and these serve to extend the reach of the technology into countless vertical markets. The different types of fingerprint products sold by these suppliers vary not only in functionality, but also in mechanical design. Essentially there are a number of main designs that are available on the market today:

- *Capacitive* – This technology is based on the measurements of the capacitance between the skin surface and the silicon sensor. The capacitance is then converted into a grayscale digital image. Another option is to inject an RF signal into the finger, in order to measure the 'live layer' under the surface of the skin.
- *Optical* – This is the most widely used sensor on the market today. The finger is placed on a coated platen and exposed to a light source. In most devices a CCD (charged coupled device) receives the image of the fingerprint with dark ridges and light valleys.

- *Thermal* – This measures the temperature differential between the skin ridges and the air caught in the fingerprint valleys.
- *Ultrasonic* – Acoustic waves are transmitted and distances are measured based on the impedance of the finger, the platen and the air. These distances are then converted into a representative image. Advantages of this type of sensor are its ability to work for long periods unattended thanks to its good performance in dirty environments. It is comparatively expensive however.

Out of these different designs, the optical sensor leads the industry, with the silicon-based sensor showing promise but always playing catch up. However, in the last two years silicon-based sensors have improved dramatically in design, and it is only a matter of time before they contend the number one spot in the industry. The other designs of sensor are only produced by a small number of players, and have a limited market size.

Numerous vendors provide fingerprint software, which they bundle with other manufacturers' sensors. Some manufacturers solely provide the sensors, making their business case by selling thousands at a time. Other providers supply both hardware and software to interested end users. In terms of market leaders, Bioscrypt, Identix, Infineon, STMicroelectronics, Authentec, Secugen and Digital Persona are among the top names.

Since September 11th, interest in biometrics has risen exponentially, and with it so has the interest in fingerprint technology. While this has been good news for the technology vendors, a number of thorns remain in their side. First is the issue of privacy, which still is a major issue, especially if the system is used for large projects by government departments. The other issue is potential fingerprint forgeries, which have been demonstrated and widely publicised in 2002.

3.3.1.1 Technical Developments

The major technical developments in this sector appear to have been made in the silicon sensor market, as manufacturers have been busy attempting to reduce the weaknesses traditionally associated with silicon sensors, such as problems with electrostatic discharge (ESD) and mechanical scratching.

Of course progress has been made across the board in the fingerprint industry, in both optical and silicon markets. Almost all vendors, for example, have continued to make their sensors available on new platforms such as Windows 2000, NT and XP and many have made their products USB compatible. The continual drive to make smaller sensors also continues.

Silicon sensors

One of the historical advantages of optical sensors are their ability to withstand difficult environmental conditions, so this was one area that had to be resolved if silicon-based sensors were to be commercially competitive. Over the last few years new coatings have been developed which form a layer of protection and the type of coating used is an important distinction between suppliers. There is still room for improvement, however, especially in damp and humid conditions and also when the sensor surface is subjected to rough handling.

One of the biggest problems was ESD – it is said that even a person building up static whilst walking across a carpeted room was too much for earlier sensors to handle. However, all the main suppliers now seem to comply with an IEC standard – IEC 61000-4-2 – with most of their products. A common figure quoted is compliance to ±15 kV and many manufacturers will aim to allow this level of ESD to be reached numerous times, rather than just once.

Another important development has been the production of sensors with ever-smaller dimensions. Thinner devices are being produced in order to squeeze them into hardware, such as smart cards and mobile phones. Meanwhile the area of the sensors is also decreasing – in particular with the style of sensor that acquires an image when a user swipes their finger across a thin strip of silicon, rather than simply pressing their finger on a defined area.

In a touch-based sensor the physical size of the fingertip governs a minimum sensor area that can be used in order for a quality image to be obtained – the lowest on the market being 42.25 mm^2. There are software tricks that can be employed to make the effective area of the smaller sensors appear larger – by stitching together a mosaic of images taken over time. However, the touch sensors cannot compete with the swipe-based sensors in terms of area, with the smallest being just under 6 mm^2.

Other than the limitations in room on a device, such as a mobile phone, another important reason for silicon-based sensors to be small is because the price of a silicon sensor is directly proportional to the amount of silicon used. So swipe-based sensors can be significantly cheaper than larger touch-based varieties.

Optical sensors

Optical sensors still rule the roost when it comes to market share in the biometric industry. The technology is robust, reliable and fairly cost effective, so currently a natural choice for end users of today's main markets. However, the main growth markets for fingerprint sensors going forward are predicted as IT security and the securing of devices, such as mobile phones.

This creates a problem for optical sensor manufacturers because their products are significantly larger today than their silicon based counterparts. In the IT security field, optical sensors can do well when presented as a peripheral device, such as an integrated mouse or a simple standalone sensor. However, integration into smaller devices, such as laptops or mobile phones is not really viable.

In scenarios, such as physical access control, however, the proven durability of the sensors is a bonus. Manufacturers have worked hard to make sure their sensors are resilient to scratching, and that they can acquire even dry finger images (which are difficult to obtain with silicon sensors). Some vendors use hard coatings on their sensors, while others use extremely hard substances for their prism surface.

Liveness testing

An area of major concern for fingerprint sensor manufacturers is the issue of spoofing. Recent studies into the vulnerability of biometric devices have shown that most sensors' security mechanisms can be bypassed with a 'gummy' finger – a fingerprint created from a material, such as gelatine or candle wax – and other clever techniques. Clearly for an industry selling its products under the 'enhanced security' banner, this is not good public relations.

A recent attack involved a researcher at the Yokohama National University in Japan, who managed to bypass 11 different sensors, and writers for a German publication, c't, who used candle wax to bypass both optical and silicon sensors.

The most resistant sensor was actually the swipe-based sensor design, as it did not suffer from simple latent image reactivation techniques, such as breathing on the sensor or dusting the surface of the sensor with graphite powder and pressing down on the resulting print with an adhesive tape.

Not surprisingly, many manufacturers defended themselves in the months that followed, often claiming that their systems weren't one of those bypassed by the attackers. Other companies have said their sensors now incorporate defences to guard against all publicised forms of attack, but most have not released any significant level of detail.

This debate is likely to rumble on in the future and has the potential to cause great harm to the industry as a whole, especially if the spoofing is done in a live environment for reasons less innocent than writing an article or research paper. It must be stressed however, that fingerprint sensors are often just one defence of many (such as a PIN or smart card) and that the attackers were in a laboratory situation with ample opportunity to try to attack the systems repeatedly. It is also a lot more difficult to get a good quality fingerprint from an unsuspecting third party than it is to use a fingerprint given voluntarily.

Smart cards

Smart cards have taken off commercially and it would make sense if fingerprint biometrics could piggyback on this success. A typical biometric/smart card combination today uses smart cards to hold a user's biometric template. This ensures that the user carries their template with them rather than it sitting on a central server or client terminal.

There are a number of possible options for matching this template against a live biometric. The simplest is for the template to leave the card's chip encrypted and for matching to take place on a remote PC – or in the smart card reader itself. The next security level up is to perform the template matching on the card itself. This ensures the cardholder's template never leaves the chip, eliminating the possibility of interception.

The highest security level would come about if the biometric sensor could be incorporated into the smart card itself. This would mean there was no longer any need for a separate reader and that all processing, including template acquisition, pre-processing and matching, would be performed on the card itself.

At present this feat is only really feasible with silicon fingerprint sensors. The difficulty still facing manufacturers is to produce a sensor that is robust enough to fit into an ISO standard thickness card (standard credit card thickness). The frontrunners in this quest include Biometric Associates, Fingerprint Cards, Fujitsu, Novacard and Infineon.

3.3.1.2 Market Developments

This flexible technology allows fingerprint suppliers to be active in most of the horizontal markets, except for surveillance. They have also found success in all the vertical markets.

In particular, the vast majority of IT security installations over the last year were based on fingerprint recognition technology – and there have been numerous health sector installations using fingerprint biometrics in the run up to the USA's HIPAA regulations which come into force in 2003.

Meanwhile some computer manufacturers have introduced PCs or laptops with fingerprint technology, such as Authentec's link up with Acer America and STMicroelectronic's connections with Samsung.

Many nationwide ID systems are based on AFIS (automated fingerprint identification system) installations. AFIS installations fall beyond the scope of this report, as they are generally massive systems where the biometric matching engine is only a small part of the overall infrastructure. (AFIS checks are also not performed in real-time, so do not fall under the true definition of a biometric system.)

Many of the large government schemes using biometrics around the world do use fingerprint technology in a true biometric sense, however. This is particularly notable with voting-applications, where a common scenario is for ID cards to be issued with a person's biometric identifier embedded within, which is subsequently checked on polling day.

New legislation will likely provide a lot of potentially lucrative contracts for fingerprint vendors. For example, the Driver's License Modernisation Act in the USA calls for the use of fingerprint templates to be used. One reason for the selection of fingerprint technology was said to be the low cost of the readers, which would have to be deployed in high numbers if the legislation was passed. Asylum seeker legislation across the globe is also backing fingerprint biometrics to facilitate their efficient processing at ports of entry and beyond.

Pricing

The price of fingerprint sensors has dropped significantly over the last two years – in particular the silicon sensors. Pricing is not an easy issue, however, as the devices fluctuate in price depending on the number ordered.

Silicon sensors are, for example, available at around US$10 for orders of 500,000 or more, some manufacturers claim. These fantastic quantities are not being realised in the industry yet, however. A more likely order size is 1000–10,000 sensors and at this level a price tag of closer to US$20–50 is more realistic.

Importantly, the price of the sensors has now come down to much more acceptable levels by manufacturers of mass consumer products, so removing one more obstacle from this technology becoming commonplace. (Mobile phone manufacturers' reluctance to include biometrics until banking applications become more widespread is a remaining obstacle.)

Although silicon technology is becoming ever more competitive in terms of price and size, optical sensors will remain competitive for some time yet. This is especially true in the larger optical reader range, which is necessary for forensic quality readings in applications, such as civilian ID. Optical sensors still perform reasonably well on price too, at well under US$100 for a typical sensor. Optical sensor robustness is also another major advantage.

Other factors

There have been a number of other developments within the fingerprint sector that could have a bearing on the industry as it develops over the next few years. One of the most significant is the announcement made in 2002 by Identix, one of the optical sensor market leaders.

Identix has pulled out of the low-cost optical sensor market and instead will adapt its verification software to also work with silicon sensors. It said that the falling price of silicon sensors, combined with their increasing stability and smaller sizes, makes them viable alternatives. This market exit for strategic reasons is a blow for optical sensor proponents.

Whether this strategic change will signal more optical sensor defections is yet to be seen. There has certainly been no obvious let up in demand for optical sensors according to manufacturers.

3.3.1.3 Drivers

- *Government acceptance* – One of the strongest drivers for fingerprint biometrics is its wide acceptance by governments. Long used by law enforcement agencies, the technology is proven, trusted and well developed. In light of September 11th, government contracts are going to prove some of the most fruitful in terms of revenue, so this acceptance will stand the technology in good stead.
- *Flexible solution* – The flexible nature of fingerprint technology extends its potential use into almost every horizontal and vertical market sector. With careful sensor selection, the technology will operate in difficult environments, whether poorly lit, dusty or humid. The size of the sensors also makes the technology a strong contender for mass markets, such as mobile phones, PDAs and laptop computers – although a time lag exists of approximately two years before this is likely to become popular.
- *Price performance* – Fingerprint sensors are highly accurate and can successfully distinguish identity from many thousand enrolees. This accuracy comes at a fairly low price, making the technology one of the best choices from a price/performance perspective.
- *IT security solution* – The dot com crash and recession have hindered the adoption of biometrics in the IT security industry. However, slightly longer term, buoyancy will return to the market and fingerprint biometrics are ideally positioned to stake a significant portion of the revenue pie.

3.3.1.4 Detractors

- *Spoofing* – The ability for attackers to forge fingerprint models that can bypass the security systems of most sensors is a serious blow. It must be addressed with robust liveness detection techniques if the technology is to be used in medium to high security applications.
- *Criminal connotations* – Unfounded as the connotation may be, there is often a reluctance for people to give their fingerprints as they fear the criminal associations of such an act. Also people do not like the idea of their fingerprint being stored on file (even though most systems store only a one-way representation of the fingerprint, rather than the image itself).
- *Failure to acquire image* – If the user base is diverse, as may be the case in wide-scale deployments, then the failure-to-enrol rate must be considered. There is always a small portion of people whose fingerprint cannot be acquired. This means that fallback procedures need to be established, which may not be desirable for some applications.
- *Physical contact with scanner required* – The fact that physical contact with the scanner is required means the sensor surface will get dirty. In an unattended application this could become problematic as the sensors rely on a relatively clean surface in order to acquire quality images. Ultrasonic and swipe sensors perform well under these circumstances.
- *Reliance on new infrastructure* – While other biometrics, such as voice or face, have relatively established networks of infrastructure (phones and cameras), fingerprint technology is starting from scratch. This means that purchasers must consider the cost of the hardware as well as the software and this may tip the balance in favour of other biometrics in some instances.
- *Stockpiled goods* – Some sensor manufacturers have reported worries that large numbers of their sensors have been purchased by IT companies, as part of contractual agreements. This was before IT markets slowed so much, and could cause a problem of oversupply in the market.

3.3.1.5 Outlook

The outlook for fingerprint technology is good. The accuracy, low cost and flexibility of the sensors are a very strong driver and this puts the biometric in pole position for the foreseeable future.

As the PC markets and mobile phone markets pick up, there will be real opportunity for mass markets to be penetrated, meanwhile government contracts will provide the strong expected shorter term growth.

In terms of revenue, the fingerprint industry was worth US$110.6 million in 2002, but this is expected to grow at an extremely healthy rate up to the end of the forecasting period approaching almost US$1 billion in 2006. At this point it will represent about 52% of the overall biometric market. The above figure assumes there is no significant relapse caused by a high profile and successful spoofing attempt on the technology.

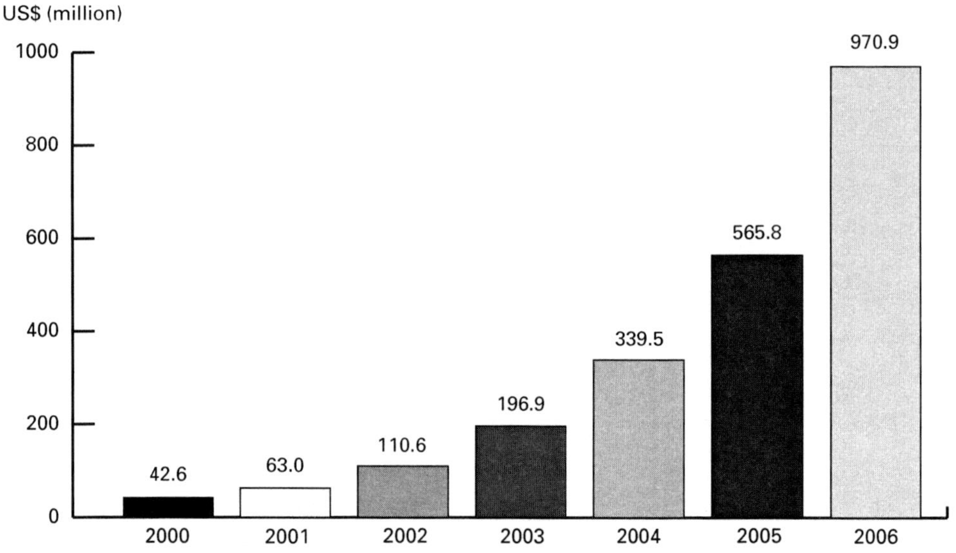

Figure 3.4 Revenue Forecast for Fingerprint Recognition Technology 2000–2006 (US$ million)

3.3.2 Hand Recognition Market

The human hand provides the source for a number of biometric techniques. Various measurements of the outer geometry of the hand have been used, extremely successfully, for a number of years. However, other information obtained from the hand is also increasingly being employed to discriminate between individuals, such as vein patterns on the back of the hand and palm prints.

Two market players, namely Biomet Partners in Switzerland, with its two-finger geometry devices, and Recognition Systems in the USA, with its full hand geometry systems, dominate the hand geometry market. In the world of vein recognition, meanwhile, there are few market players with developed products, although Vein ID in the USA and Neusciences in the UK are the most prominent players.

Palmprint recognition seems to have been developed mainly by livescan system providers, such as Cogent Systems, Heimann Biometric Systems and Identix. Although perfectly possible, it is uncertain whether any palmprint products are yet used for real-time authentication (biometric systems in the true sense of the definition). They are typically being used in criminal applications instead, with palmprints from crime scenes being checked for possible matches.

The geometry-based systems are primarily used as physical access control/time and attendance devices. They are typically rugged machines, capable of being installed in harsh environments and are generally cost effective and easy-to-use biometrics.

3.3.2.1 Technical Developments

Geometry systems

One of the greatest strengths of hand recognition devices is their ease of use. In almost all the geometry and vein-based systems a user's hand is placed palm-down on the device for a brief period to allow one-to-one verification or even one-to-many identification to take place with relatively small user populations. In some devices, guiding pegs are used to direct the user's fingers into the right position.

In Recognition Systems' product, a solid state camera captures a TV-like image of the top of the hand, with the added use of a mirror. It then measures the 3D size and shape of the fingers and knuckles and compresses that information to form a nine byte template – over 90 measurements are taken. Due to the standard positioning of the device's guiding pegs, only the right hand of a person is normally measured. Although, with practice, a person's left hand turned palm upwards can be used. (There have been certain one-off installations using readers that accept left hands, but this is not usual.)

Biomet Partners' system only looks at the shape and characteristics of the index and middle fingers of either hand and has a template size of 20 bytes. Users make a "v" shape with the first two fingers of either hand and place these on a platen. The user is assisted with audio, visual and tactile prompts to allow correct finger positioning. Verification time is reportedly within one second.

In both systems, a PIN must be entered in verification mode, in order to call up the user's reference template. This PIN can either be typed into a PIN pad or increasingly it may be stored on a smart/magstripe card.

Due to the ease of use of most hand-based geometry systems, they do not tend to suffer from high Failure to Enrol rates, which can be an issue with other biometric systems. However, factors such as weather, temperature, certain medical conditions and the process of ageing (especially in the very old and very young) can sometimes alter hand size. In an application where the system is well used, and there are frequent template updates over time, good FAR/FRR statistics can be achieved. Software developments over the last few years have improved the ability of systems to take into consideration items such as rings and cosmetic fingernails.

The size of the devices, which limits their use in some applications, is not an area that has been actively addressed in recent years. A recent technology addition to Recognition System's product range, however, is a system that combines hand geometry and facial recognition. At five seconds, it is slightly slower than the hand geometry system when verifying a person's identity, but it does reportedly enhance the performance of the device.

Vein systems

Vein systems generally work by recording an infrared image in the back of the hand, providing a large stable image. They have not yet had a great deal of commercial success, but do have some technical advantages that set them apart from other biometrics. For example, vein patterns are generally hidden and therefore much more difficult to forge compared with a standard hand geometry system. The veins are also well protected from damage.

Liveness detection – a major issue in the biometric industry at present – could be installed fairly easily with vein-based systems, which measure the oxygenated blood flow that gives rise to the vein patterns. Template size can be as low as 300 bytes and in some systems can be held on a smart card by the user.

3.3.2.2 Market Developments

Hand-based geometry technology can attribute the majority of its success to the physical access control and time/attendance horizontal markets. However, the citizen identification market has also borne contracts for suppliers, particularly in verifying the identity of travellers at immigration points.

Due to the diverse number of locations where these kinds of devices can be found – basically anywhere where there is a need to secure a door in a reasonably secure and user-friendly way – there is no single dominant vertical market. However, the law and order, immigration, industrial and leisure industries have secured the most high profile orders, to date. That said, everywhere from post offices and bus fleets to hospitals and sports stadiums have installed the readers.

According to Recognition Systems, the clear market leader, it has sold in excess of 55,000 of its HandReader units worldwide. The majority of these have been installed in small numbers, but there have been some significant deals in its history including:

- The 1996 Olympic Games, where HandReaders were used to protect access to the Olympic Village. More than 65,000 people were enrolled and over 1 million transactions were handled in 28 days.
- 900 units control client and employee access to special areas of Italian banks.
- The Immigration and Naturalisation Service Passenger Accelerated Service System (INSPASS) is installed at many US airports to allow many thousands of enrolled passengers to pass quickly through immigration. Figures from earlier this year said that there were 29 kiosks in operation, with over 90,000 people enrolled in the system. Between December 1999 and November 2000 some 280,000 verifications have reportedly taken place.
- Ben Gurion Airport, Israel, exceeds the INSPASS numbers with over 100,000 people enrolled in the first year alone, growing at approximately 2% per month. Around 50,000 verifications are performed each month.
- The United States Federal Bureau of Prisons and UK prisons rely on the units for prisoner tracking.
- Still not up and running is the technically-challenging Basel border crossing project in Israel, where hand geometry and facial recognition will be combined to confirm the identities of tens of thousands of Palestinians crossing to work from the Gaza Strip to the West Bank.

Biomet Partners also has some high profile implementations, the most notable of which was the use of its technology in a number of Disney parks to oversee admittance of season ticket holders. Other projects include:

- 2,000 Italian post offices will use the technology for employee 'mantrap' entry.

- A Brazilian bus fleet operator is using it to control use of season tickets and special passes.
- Various contract wins with the US Department of Defence to protect weapons facilities, control prisoners and secure computer facilities.

As a technology, hand-based recognition has the best track record for the sheer number of successful implementations. Time and attendance is a very well received add-on application to hand-based geometry systems, and this is likely to be one of the areas of strongest growth in the future. It provides a clear payback for the system owner, in particular because it can help prevent 'buddy punching', which can be a problem in large work places. This is important because the price of hand geometry devices is at the high end of the access control spectrum with a typical installation costing from US$550 to US$4000 depending upon the configuration.

3.3.2.3 Drivers

- *Track record* – Hand and finger geometry systems have a track record unparalleled in the biometrics industry. Although this issue is a major detractor for some other biometric technologies, geometry systems have been there and done it. This gives confidence to potential purchasers and plenty of ammunition to sales staff to help secure contracts. This driver does not apply to vein-based systems, which have an almost zero track record.
- *Easy to use* – The ease of use of hand based systems is clear. Very little training is required for users and there are often guidance pegs or audio/visual prompts to ensure the hand is positioned correctly. In large populations of people, such as you might find in a factory or prison setting, this has a clear advantage over biometrics, such as iris or fingerprint recognition. The only issues may arise with old people, particularly those with arthritic fingers that do not spread easily.
- *Good user acceptance* – Hand biometrics does not suffer from privacy invasive connotations found with some other biometrics. Combined with its ease of use, there is typically strong user acceptance for this biometric.
- *Rugged design* – Hand and finger geometry systems are perfect for operation in hard environmental environments. Lighting is not a significant issue, neither is dust or dirt.
- *Fraud resistance* – If a person willingly gives up a cast of their hand or fingers then it could be possible to spoof a geometry-based system. However, this is highly unlikely and other methods would rarely be worth the effort bearing in mind the typical scenarios in which the readers are used.
- *Small template size* – At just nine and 20 bytes respectively, hand and finger geometry systems have the lowest template sizes in the business. This makes it easy for standalone devices to hold many templates or for individual templates to be stored on cheap, but low capacity, magnetic stripe cards.

3.3.2.4 Detractors

- *Accuracy issues* – hand geometry systems cannot perform 1:n searches with large populations of people, due to the relatively low data set that the systems have to work with. This limits the use of the system to certain types of scenario. Also, if people do not use the system on a regular basis then the system can experience higher FRR due to the natural variations in the shape/size of people's hands over time. As mentioned earlier, the hand changes under certain environmental conditions, as well as when a person is growing up (until after the age of eight approximately).

- *Physical design limitations* – The fairly large, rugged design of hand based geometry systems is a driver when looking at environmental conditions, but it is also a detractor as it eliminates hand-based recognition from many up-and-coming biometric markets, such as IT security.
- *Cost* – The price of hand-based recognition systems is generally quite high compared with other less technologically advanced systems. However, compared with other biometrics, such as iris recognition, the system is cost effective. That said it may not provide the level of security or the 1:n hands-free matching capabilities of other biometric systems. A well-installed time/attendance system may help alleviate the cost issue, through a clear payback structure.
- *Hygiene issues* – Although rare, there are reports that some nationalities of people, such as the Japanese, do not like putting their hands on a surface straight after others. Obviously, if the surface gets extremely dirty, this may also be a put off for people using the system.

3.3.2.5 Outlook

Hand-based biometric technology is a strong performer in today's biometric market. This strong order base is expected to continue to accelerate, but over time the market share of the technology will slip, as other biometrics push forward into high-growth markets.

Along with fingerprint biometrics, hand and finger geometry technologies will continue to be leaders in the physical access and time/attendance market, but in the absence of any mass market potential, overall growth will be moderate.

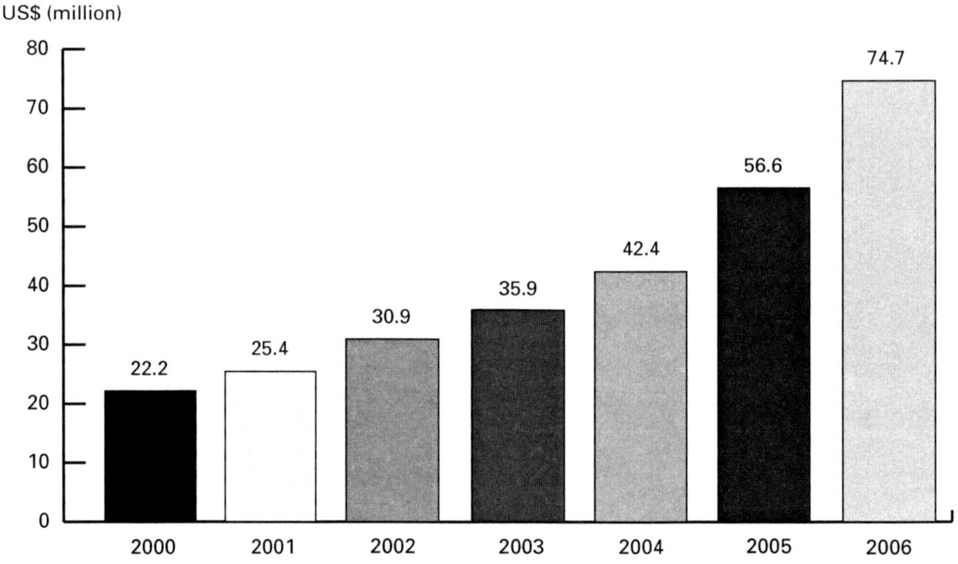

Figure 3.5 Revenue Forecast for Hand Recognition Technology 2000–2006 (US$ million)

In revenue terms the technology represents approximately 12% of the market in 2002 at just under US$31 million. This is expected to fall to 4% of the market by 2006 with a revenue of US$74.7 million.

3.3.3 Facial Recognition Market

Faces are probably the most natural vehicles for carrying and delivering identity information. But although almost any camera can be used to obtain an image of the face, the technological hurdles in automating the processes of face recognition are still formidable. The benefits of a successful facial recognition system are great, however, and that is why, despite the challenges involved, face verification and identification technologies are fertile grounds for the growth of biometric applications.

There are numerous techniques employed in order to perform facial recognition, including eigenfaces, local feature analysis, automatic face processing, 3D analysis and neural networks (see Table 3.2). Unsurprisingly, the key to an effective facial recognition system is the ability to work with quality facial images, both at the enrolment and subsequent live image-capturing phase. The best-case scenario would have good lighting, co-operative users and correct camera and user positioning. Unfortunately this set of conditions is often not possible, especially in surveillance applications or where the system has to fit into pre-designated locations (ie at ATM points or an existing CCTV camera network), so accuracy levels are often impacted negatively.

Another problem is that the face is the most variable part of the body, thanks to a multitude of facial expressions, beards, changing hair styles and the use of make up and headwear.

In the wake of September 11th, facial recognition companies' stock prices soared as it was perceived that this technology could provide a way to look out for criminals in airport environments. However, with this new demand for the technology came all the negative connotations associated with covert monitoring systems. This has meant that out of all biometrics, facial recognition technology has proven to be the most controversial in recent times, especially from a privacy perspective.

Since its high profile billing, the accuracy of such systems in certain environments has also been questioned. In the months that followed September 11th, numerous tests have shown facial recognition technology to be less than perfect in surveillance applications. Nevertheless, interest remains high. After all, if you can catch twice as many undesirable people compared with before then that is an improvement that cannot be ignored. (Note: phrased this way, rather than saying the system still lets 'x' percent of the undesirable people through, which is the way performance has been quoted by most anti-face recognition groups.)

Facial recognition technology has so far found it difficult to penetrate the IT security market, but its use in access control applications and forensic-type applications is still an area of interest.

Table 3.2 Typical Facial Recognition Techniques

Recognition technique	Explanation
Eigenfaces	This technology was developed at MIT in the USA. It works on the assumption that faces can be categorised according to their 'degree of fit' against a number of master eigenfaces.
Local feature analysis	This technique is more accommodating of changes in appearance or facial orientation. Numerous features from different areas of the face are analysed, as well as the relative location of these features. As the technique does not rely on seeing the whole face, horizontal and vertical angles of between 15–25% can be catered for.
Automatic face processing	This simplistic technique looks at distances and ratios between features of the face that are relatively easily acquired, such as the eyes, nose and corners of the mouth.
3D analysis	This technique is mostly at the R&D stage of development, but involves constructing three-dimensional face models. The use of structured light to illuminate the face with a known regular intensity pattern is proving most popular. Conventional cameras can then be used to measure the reflected pattern and convert the spatial distribution of the received patterns into depth information, from which the face model can be constructed.
Neural networks	Neural networks use an algorithm to determine the similarity between unique features of live versus enrolled faces. Each feature is analysed and a 'yes' or 'no' decision on a match is given. In the event of a false matching decision from any particular feature, the matching algorithm reduces the weight given to that feature for future decisions. Over time, this should help the algorithm work effectively in difficult environments.

3.3.3.1 Technical Developments

As world interest in facial recognition technology grows, so the suppliers in the industry are quickly having to look at and resolve technical issues that still need addressing. Among the most important are image acquisition capability, technology downsizing, accuracy and spoofing.

Image acquisition

Failure to acquire rate is a significant variable for facial recognition, especially in surveillance applications. One well-known industry player once said: "We cannot perform magic. If there's no face we cannot acquire it."

There are a number of ways that could be used to improve the acquisition rate however. In a verification scenario, for example, machine feedback is one option with the device actually prompting the user via on-screen instructions or an audio cue. Alternatively, the addition of an LED or mirror is an option to coax the user to look at a certain position. Increasing the field of view of the camera can also be beneficial.

In surveillance applications, the intelligent positioning of cameras is one of the most obvious ways to ensure better image capture – for example by putting a clock or display on a wall facing passengers as they descend an escalator. Other options in completely unconstrained environments, such as you might find in a public place, is the positioning of sonic aids which have the ability to make a person look up at the camera.

A problem with capturing faces of people moving around is that the size of the head will increase or decrease depending upon whether the person is getting closer or further away from the camera. Intelligent camera systems are now available which will lock onto that person's face and then adjust itself to keep the head size constant, so enhancing the chances of acquiring a useful image. The correct environmental conditions are also important and wherever possible system integrators attempt to position cameras in a well-lit, uncluttered environment.

Downsizing existing technology

The widespread availability of cheap digital cameras, often bundled with desktop computers and the increasing appearance of mobile phones with image capturing capabilities, are positive trends for facial recognition vendors. However, while the widespread availability of sensors looks likely, the nature of the hardware and its limitations will produce new challenges for software designers.

As well as the inevitably difficult environmental conditions, some of the challenges of cheaper sensors include more noise and limited capabilities of small LCD displays. Processing power will also become an issue if sophisticated facial recognition algorithms are to move into these handheld devices.

Power consumption is another issue that must be resolved when it comes to downsizing technology. Some suppliers are investigating the possibility of creating purpose built chips that embed a face recognition instruction set, so helping to reduce the amount of processing power required.

Accuracy

The accuracy of facial recognition engines has been under increased scrutiny since September 11th. Following the terrorist attacks, interest in using the technology as a tool for surveillance monitoring has increased and a number of trials have been performed at airports in the USA, such as Palm Beach International; Boston Logan International; Dallas/Fort Worth International; St Petersburg-Clearwater International; and Fresno Yosemite International.

In particular, accuracy levels have been criticised by privacy groups. In a recent news item, the American Civil Liberties Union (ACLU) said that results from

Palm Beach International Airport showed the failings of the technology because the technology had failed to pick up around 50% of volunteers on a watch list.

These statistics were misleading because the trials at that particular airport were attempting to virtually eliminate false alarm rates. A fairer assessment of the technology (according to suppliers) is when the sensitivity of the system is set to medium. At Dallas/Fort Worth International Airport, for example, the correct alarm rate was around 94% while the false alarm rate was 1.5%.

Recent trials at Virginia Beach using Identix technology also passed the customer's performance criteria.

In verification mode it is reasonable to assume a cooperative audience and therefore high image acquisition quality and rate and subsequent system accuracy. However in a surveillance situation, subjects wishing to be avoided are likely to be completely uncooperative – taking evasive measures or by using disguises. This would almost certainly render the system less effective. The location is important to bear in mind, however. Clearly an uncooperative user would stand out in an airport environment and warrant further investigation anyway.

Spoofing

There have been recent reports suggesting that facial recognition systems can be easily circumvented using photographs. This is a serious problem and some vendors claim the 'liveness' tests within their systems can protect against forged faces. Such a test can be based on detecting movement that fits a model of plausible facial motion.

Another possible attack may come from recorded videos that may pass ordinary liveness tests. Here challenge-response techniques are being investigated, where the system may prompt the user to make specific facial gestures or to say specific words or phrases prompted by the system. Joint audio-visual systems make forgery attempts much more challenging. Also the application scenario must be considered. In an access control scenario at a factory, for example, it would be extremely difficult to spoof the system with a monitor based attack method.

3.3.3.2 Market Developments

Out of all the technologies in the biometric industry, the facial recognition market has provided some of the most interesting headlines over the last year. The technology is making strong headway in surveillance, physical access control and citizen ID horizontal markets. PC/Network access applications are lagging behind, in the face of a subdued market and strong competition from fingerprint technologies.

In terms of vertical markets, the travel/immigration, government and law and order markets are proving to be the best performers, followed by financial services. This is unsurprising in light of the events of September 11th, where the focus on use of the technology in airports and in national ID/passport schemes intensified.

The main contract wins over the last year can be broken down into the various horizontal and vertical market categories, such as surveillance, citizen ID, law and order, travel and immigration, leisure and forensic type applications.

A brief selection of installations included:

- Public surveillance system at Virginia Beach;
- Numerous airports, such as Sydney International and the selected US installations listed above;
- Public surveillance in Tampa, Florida, in the downtown district of Ybor City;
- Driver's license contract wins in the US including the state of Connecticut and West Virginia;
- Numerous contract wins at casinos worldwide, including: Harrah's Joliet Casino and Hotel in Chicago and an unnamed South Korean casino;
- Police department projects for accessing mugshots, such as the BlueBear project at police departments across Canada and a central picture system for police officers in Belgium;
- Uganda's Electoral Commission's system to de-dupe photographs of freshly enrolled citizens (The goal was to reduce voter fraud in the country, which was believed to be rife);
- The s-Travel initiative, funded by the European Commission and the Swiss Office for Education and Science, will use facial recognition technology combined with smart cards and digital certificates to enable secure physical and electronic airline and airport processing for frequent air travellers. The trials are taking place in Zurich and Milan, with a view to expand globally.

Away from the pure contract side of the market, the technology is influenced by other factors, such as privacy concerns, legislation and appropriate revenue models.

Privacy groups have been a constant thorn in the side of the industry and up until now, privacy concerns are still unresolved – this is despite the efforts by the main industry players, and more recently the legislators, to set guidelines for the use of the technology in public places.

As an example, an anti-facial recognition report was released by the ACLU, a strong US privacy group, condemning the use of facial recognition to monitor people in public places. It specifically pointed to the Tampa installation in Florida, where police started scanning faces in June 2001 in the Ybor City Neighbourhood.

The report was damning, although, as may be expected, was strongly contested by the industry. Nevertheless it highlights that facial recognition systems, in particular surveillance systems, have a long and rocky road ahead. Clear guidelines for the systems' use are being adopted but this is unlikely to appease staunch privacy advocates and silence their negative sound bytes.

A very positive development in the market is the approval of various legislative initiatives in the USA, which have come about in response to September 11th. The most prominent for the facial recognition market is the *Enhanced Border Security and Visa Entry Reform Act of 2001 (S.1749)*, which requires all travel documents issued after October 2004 to contain a biometric feature. Other legislation surrounding driver license issuance is also under consideration.

The Visa legislation states that countries wishing to take part in a visa-waiver programme with the USA, must have a scheme in place that issues their nationals with tamper-proof machine-readable passports incorporating biometric identifiers that comply with standards established by the International Civil Aviation Organisation (ICAO). Although the choice of biometric identifier is not set in stone, an assessment by ICAO established that face recognition was the single most likely biometric to be combined with machine readable travel documents in order to allow machine-assisted ID confirmation of passengers.

As over 500 million MRTDs are in circulation globally, a number that is increasing rapidly, this represents a significant market opportunity for biometrics. The group looked at a series of weightings that took into account factors, such as enrolment, renewal, machine-assisted identity verification requirements, redundancy, public perception, storage requirements and performance. Facial recognition scored highest at over 85%, followed by finger and eye-based recognition at around 60–70%. Signature verification, hand geometry and voice verification scored under 50%.

Despite the many positive projects detailed above, there have been some bad news stories, which demonstrate the fragility of the facial recognition market and the necessity for suppliers to ensure their revenue generation models are sound.

Towards the end of 2001, the US cheque cashing company Innoventry closed its doors, due to difficult economic conditions. The company had enrolled 2.5 million people, cashed 8 million cheques and installed 1400 machines. The problem was that each machine cost US$80,000, but only made a small amount of money per transaction (believed to be around US$0.05). This meant it needed a large number of transactions to be successful.

At the time, the supplier said a healthy US economy would have created the transaction levels needed, but that the economic downturn had caused usage of the service to drop. Clearly if a model which uses existing infrastructure can be found, such as using ATMs or laptops with cameras, then a more viable financial model would be achievable.

From a competitive viewpoint, the main companies in this field include Identix (formerly Visionics), Viisage, Imagis Technologies, Acsys Biometrics, Cognitec, ZN and so on, although there are many other companies heavily involved in this field (see Chapter 5).

The highest profile installations have been by Identix and Viisage, which are the two companies vying for market supremacy.

3.3.3.3 Drivers

- *September 11th* – The September 2001 attacks caused a massive increase in the awareness of biometrics. The hope was that they could be used as modern tools in the fight against terrorism, particularly in airports. Facial recognition is probably the biometric that received the greatest initial consideration, as it could be deployed in surveillance-type installations and to check passport images. (Some facial images of suspected terrorists are available, while their other biometrics are not.)

- *Use in surveillance* – Following on from the above point, the use of biometrics in surveillance applications is a strong driver. No other biometric is as successful in recognising people from a distance as facial recognition. There have been a number of installations since the attacks and many more are expected in the short to medium term. Surveillance elsewhere, such as in casinos, and on the streets of a handful of cities in the UK and USA, are also providing good results, which may increase adoption rates elsewhere.
- *Legislation* – A strong driver is the raft of legislation, which has been drafted in since September 11th. This covers many different areas, such as visa reform and driver licenses. Facial recognition technology stands a fairly good chance of being implemented as a necessity thanks to the wording of the legislation – see Market Developments above.
- *Legacy image use* – If there are good quality images available, then it is possible that photographs of people could be used without their having to enrol. Other than fingerprints and static signatures, most other biometrics have no legacy databases that could be used.
- *Non invasive* – Aside from privacy issues, facial recognition technology is generally perceived as non-invasive, as a person must just look at a camera for their biometric to be acquired.
- *Large company backing* – Some of the biggest companies in the biometrics industry, such as Identix and Viisage, have put their financial muscle behind facial recognition and have case histories to back them up. It is less likely that a government department, for example, is going to put its faith in a system sold by a small company with little track record.

3.3.3.4 Detractors

- *Privacy* – The concerns of privacy advocates and, to a certain extent, the general public, that facial recognition could be used to monitor people without their knowledge and against their will is a strong detractor for the facial recognition industry. This is because many of the uses of facial recognition in surveillance applications do involve monitoring the public, and purchasers of such systems are reluctant to deal with the negative publicity. As time goes on, and there are more systems successfully installed, this detractor will diminish, however.
- *Performance and expectations* – There has been much written about the doubtful performance of facial recognition systems. While it is true that it is far from perfect, it can be a powerful tool if implemented properly. Until the expectations of facial recognition system purchasers are lowered (by not over-hyping the technology in the first place and through honest education and literature) many will ultimately be disappointed with the performance of the technology.
- *Easy to fool* – There are reports that the technology can be easy to spoof or evade. Anything from wigs, false moustaches and hats could make surveillance difficult. In one-to-one verification scenarios, there have been reports of the systems being easily fooled with photographs and video footage. This may prove to be a detractor for some higher security and unattended installations.

- *Reluctance to have picture taken* – Some people do not like their picture to be taken. Clearly this is a problem in some applications, which must be considered.

3.3.3.5 Outlook

The facial recognition market is one of the most promising of all the biometric technologies. The surge in interest in surveillance technology will serve to increase revenues, while the potential for the use of biometrics in passports is also a major potential area of revenue growth.

Clearly access control will continue to provide a level of revenue, as will its use in forensic-type applications or mug shot systems used by the police.

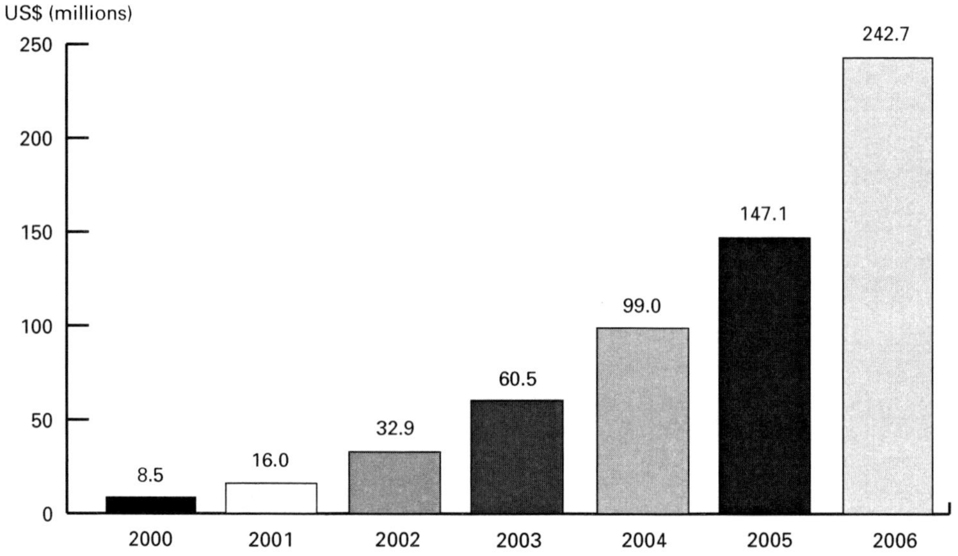

Figure 3.6 Revenue Forecast for Facial Recognition Technology 2000–2006 (US$ million)

Overall the revenue for facial recognition is expected to increase from a base of US$32.9 million in 2002 to US$242.7 million and a market share of 13% by 2006.

3.3.4 Speaker Verification Market

The creation of computer-based systems that analyse a person's speech patterns to provide positive authentication has been under serious development for more than a decade. So it is surprising that until recently only a handful of reference implementations have been in evidence to prove the validity of such technology in the commercial world.

Fundamentally speaker verification and identification technologies have it all to play for. After all, the systems provide one of the simplest and most natural modes of authentication and they can operate without the need for specialised equipment – a microphone is a standard accessory of any multimedia computer and verification can also be performed over standard phone lines.

The problem has historically been one of robustness and reliability limitations. This situation is gradually getting better, however. Microphone technology is constantly improving, while the cost of the sensors is comparatively inexpensive. Meanwhile, the processing power needed to analyse the voice can be packed into increasingly small, low power devices, and as signal processing technology evolves, new and efficient algorithms for extracting and compressing the key features of a speech signal are emerging.

A person's voice is a natural indicator of identity and the principle of a speaker-based biometric system is to analyse the characteristics of that voice in order to determine whether they are sufficiently similar to characteristics stored on file as a biometric template. This process can be either in a one-to-one verification scenario, or in a one-to-many identification configuration.

Generally speaking there are three main ways that people can use a speaker verification system:

- *Text dependent* – This requires the user to utter a pre-determined word or phrase. Most vendors offer this type of authentication mechanism.
- *Text prompted* – This requires the user to utter random words or phrases from a pre-enrolled set. This is popular in corrections monitoring as it can guard against tape recordings of a person's voice.
- *Text independent* – This type of system lets the user speak freely and can be very unobtrusive if the speech is verified whilst in conversation with a call centre operator. The user enrols in the system by saying a passage of text that contains several examples of all the phonemes that characterise a human voice. When a user subsequently verifies their identity they will normally need to provide a passage of speech long enough to contain clear examples of these phonemes (usually around half a minute's worth). Obviously the more speech provided the better the probability of a correct matching decision.

No single supplier dominates the voice biometric landscape. Some suppliers use other companies' engines in their solution offerings, while some provide the entire solution using their own verification engines. Others focus on providing the verification engines alone. Broadly speaking the landscape could be separated into solution providers and technology suppliers.

3.3.4.1 Technical Developments

All voice-based biometric systems require a user interface, software or hybrid hardware/software to convert digital speech data into templates, as well as a matching engine, which allows speech templates to be compared to one another.

There are many things a developer can do to enhance the voice signal from the user interface, which might only be capable of giving a sub-optimal voice signal. The sort of problem a vendor might have to put up with includes background and random noise, poor quality signal channels and signal coding treatments. These sorts of problem have to be considered when designing the system – so a system that operates primarily on a mobile phone network will have to look specifically at signal coding treatment, for example.

Architectures

An important choice for companies wishing to install speaker verification technology is where to place the system components in a distributed context. Some suppliers can perform the matching processes within a standalone device, such as an embedded system, while others use client server configurations, where authentication is provided remotely.

Standalone configurations could include a simple PC/microphone scenario, but it also includes embedded solutions in devices such as a mobile phone. A major advantage in this architecture is that the voice data does not leave the phone, although processing power is an issue.

Standalone solutions are not generally feasible using standard phone technology, as the addition of software to the phone handset is generally unfeasible in the majority of landline phones on the market. Such systems are best served by a server-based approach – although more severe degradation issues must be faced in this scenario.

Feature extraction

The conversion of digitised speech data into representative templates relies on some form of feature extraction process. These generally look at pitch and formant frequencies found in the voice.

The feature extraction process, while similar to that for isolated word recognition, is mainly aimed at describing and preserving information that characterises the way a person speaks rather than the words that are delivered. This process is best achieved by looking at the higher frequencies within the bandwidth.

Public telephone network infrastructures comply with a base bandwidth specification of 300Hz to 4300 Hz. This could potentially be a problem for female speakers, who often have characteristics above this upper range. However to date no serious drop in accuracy has been noted.

Clearly the amount of information extracted is proportional to the size of the speech template. Typical template sizes range anywhere from 40 bytes to tens of kilobytes. Research is still underway to try and reduce template size without compromising quality and therefore performance. The benefits of small template sizes include speedier transmission across telecommunication networks and faster database searches if the technology is used in an identification configuration.

Performance

While fingerprint, hand and face recognition biometrics are leading the way in terms of systems deployed, voice biometrics perform surprisingly well in tests. In the UK, the Communications and Electronics Security Group (CESG) carried out tests on seven biometric technologies. These showed the voice-based systems to be second to iris recognition in overall performance (where three attempts were given to access the system).

Although the performance situation is improving, it is still important for potential users to do the maths when considering rolling the system out to the population at large – in a mobile commerce scenario, for example. Even at low equal error rates (EER) a large number of users would still be falsely rejected on a regular basis and back up procedures would be necessary.

Performance can be influenced in many ways, but one important influence is that of the user interface itself. Allowing the user to speak in as natural a fashion as possible can help as it helps to reduce stress and avoids speech variations. One factor which causes increased stress in the voice is a poorly designed interactive voice response (IVR) system that leaves a caller on hold or having to wade through many levels of options. If the user has to utter too much text, this can also cause problems, as the pitch or syllable stress may vary.

Spoofing

Fooling a biometric system is possible regardless of the biometric being used – as evidenced by recent spoofing attempts, which even cracked iris recognition.

With speaker verification, the worst case scenario is if somebody replays a recording of a person's voice into a microphone. Although there are some voice systems that claim to discriminate between live and recorded speech, they are by no means infallible.

However, in speech technologies' favour is the ability for the system to use operational techniques that make forgery or replay attack more difficult. The simplest is to present a random subset of words taken from a full set of words taken at enrolment. However, other techniques, such as challenge/response systems can also be used.

3.3.4.2 Market Developments

Voice based biometric systems are making headway in many of the horizontal markets but the main ones include PC/Network access and financial transaction authentication. The vertical markets where speaker verification is proving most popular are the law and order, telecoms, financial services, healthcare, government and commercial sectors.

Notable contract wins over the last year have been the launch of an account access application by Banco Bradesco in Brazil; the deployment of a call centre (for PIN/Password reset) operation for the publisher Associated Mediabase; the tracking of young offenders in the community by the UK's Home Office and a system which allows Union Pacific Railroad customers to arrange the return of empty railcars once the freight has been delivered.

Major applications

The use of speaker verification (sv) technology in law enforcement applications is one of the biggest sectors for the industry. In particular, the use of the technology to monitor parolees or young offenders (released into the community) is proving successful, with numerous roll outs in European markets, in particular. Another successful application in this sector is the monitoring of inmates.

Call centre-based applications are proving strong hunting grounds for suppliers of voice systems. A trend that is evident in this area is the combination of speech recognition and speaker verification systems in what is often referred to as a total voice solution. The advantage is that it provides customers with a 24-hour automated service, which can extend the application further than a speech recognition configuration alone. This is because speaker verification adds security to the system and so allows applications to bypass the need for a caller to speak to a call centre operator to verify identity.

The use of speaker verification in password reset operations are also proving popular. It is often reported that the most common call to a company's help desk is because of forgotten passwords. This is an ideal application for speaker verification, which can be used to authenticate the identity of the employee and then automatically reissue a PIN or password to the caller.

Looking forward, the possible use of speaker verification technology in mobile phones or new PDA technology provides an exciting opportunity. As the third generation phones are launched over the next year, it will quickly become apparent that their functionality also brings a security headache.

This is one area where sv technology can provide an answer – although it faces hot competition from other biometrics, such as fingerprint, dynamic signature verification and iris recognition. Nevertheless there are companies that have made strong progress in the mobile phone sector, such as Domain Dynamics in the UK, which recently demonstrated the possibility of using a SIM card to perform authentication (it did this in conjunction with Schlumberger Sema and Mitsubishi).

Corporate news

From a corporate perspective an important development has been the number of integrators and value added resellers, which have now appeared on the market. This helps to spread the reach of voice biometrics into new vertical markets and new geographic locations.

Elsewhere changes have included the sale of SpeakEZ Voice Print technology from T-Netix to Speechworks International and the closing down of leading supplier Lernaut & Hauspie Speech Products – which sold its business and technology assets to ScanSoft.

In terms of technology providers, a selection of the market leaders include Nuance, VoiceVault, Speechworks, Domain Dynamics, Veritel and Persay.

3.3.4.3 Drivers

- *Existing infrastructure* – Most biometrics providers have to worry about providing hardware to end users of their technology. Speech based systems are in the privileged position of already having a huge infrastructure in place that is capable of supporting its requirements – namely the landline and mobile phone networks.

- *Improving accuracy* – This was once a significant problem for voice biometrics. Today this is improving as evidenced by tests carried out by CESG in the UK. However, in the type of application that promises the greatest future growth potential – e- and m-commerce – the accuracy will still have to improve unless many legitimate users are turned away.
- *Increased number of VARs/integrators* – the increased number of VARs and integrators will bring speech-based biometrics to the market in applications so far untapped.
- *Call centre automation* – Increasingly the use of sv technology to process callers to call centres is being introduced. This trend is likely to continue as the need to give customers greater functionality increases along with the continual pressure to cut call centre operation costs.
- *m-commerce opportunities* – the emergence of commerce opportunities on mobile phones will provide a future growth opportunity for sv technology. The technology could also be used to cut down on mobile phone theft – as phones that only respond to authenticated users emerge.

3.3.4.4 Detractors

- *Accuracy* – As detailed above, the accuracy of speaker verification technology must still be improved in order for the greatest potential markets to be tapped. In particular accuracy is dependent on establishing clever user interfaces and systems to enable the user to speak as naturally as possible.
- *Competing technologies* – The speaker verification market faces stiff competition in markets such as m-commerce from other biometrics, such as fingerprint, and verification techniques, such as PINs and passwords.
- *Slow adoption* – Apart from a few exceptions, mobile phone operators have been slow to answer the security challenge, instead focusing on other challenges such as the roll out of new generation networks, increased phone functionality, bluetooth and so on.

3.3.4.5 Outlook

No other biometric has the advantage of an infrastructure quite as established as that of voice biometric technology and this could prove to be a winning formula over the years to come. Now that accuracy and robustness problems appear to be resolved, there is every chance this biometric will do well. In particular, call centre applications are expected to provide a great boost in short-term growth, with e- and m-commerce opportunities emerging in the next three years.

From a revenue perspective, in 2002 the amount of money created by the industry stood at US$13.9 million, some 5% of the total market. This is expected to keep at a similar market share to 2006, which is impressive considering the strong growth of other technologies such as fingerprint, face and iris recognition.

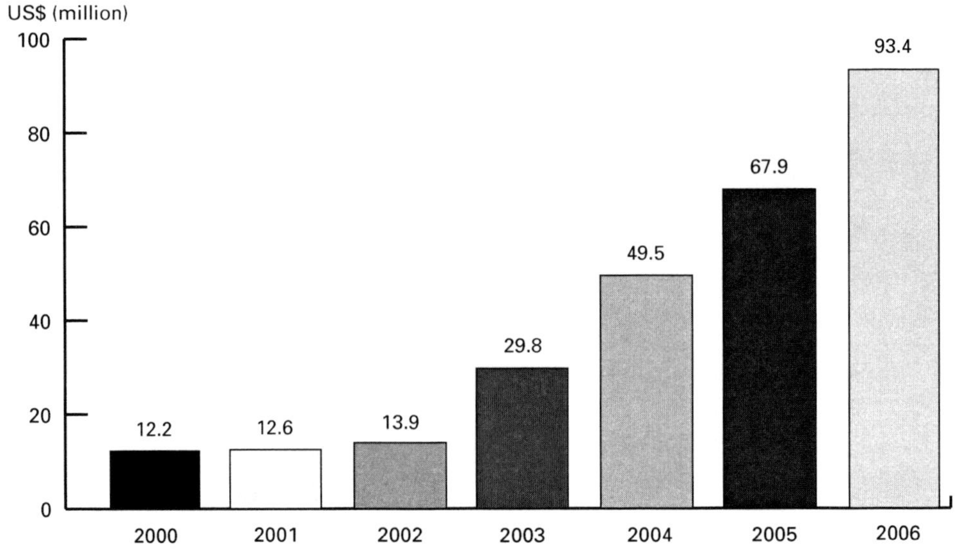

Figure 3.7 Revenue Forecast for Voice Verification Technology 2000–2006 (US$ million)

3.3.5 Signature Recognition Market

Signatures are a natural, traditional and accepted way for people to prove their authenticity, whether it is as part of a financial transaction, in the preparation of a will or the signing of a business contract. A person's signature implies their agreement with the transaction or business deal and creates a bond that can later be submitted as evidence in a court of law.

Although a signature is a good identifier, it can be forged however, and is especially vulnerable when it is not possible to check a person's other credentials. For example, when they are required to sign a document, such as a credit card slip, in a busy supermarket queue.

So the problem is being able to verify that the person making the transaction is actually the person they claim to be. Security levels can be raised, by insisting the person gives a PIN or a password, but in the end the question of true identity is still left unanswered. This is where biometrics, such as dynamic signature verification (dsv), are able to make their case. When a person's biometric is authenticated, the chance of an impostor getting through is greatly reduced.

Another case for dsv technology can be made thanks to the increasingly paperless society we live in today. Traditional paper and ink approaches would clearly no longer be viable when deals are being conducted over the internet or when a company wants to eliminate its paper burden and store all its documents electronically. This is where a dynamic biometric template can come into its own. Embedding a fresh biometric template of the signatory into the electronic document, gives irrefutable proof that the right person signed the document and the signatory cannot later refute that act.

In essence, using dsv technology brings back the confidence given by a pen and paper solution by combining identity verification with intent in a traditionally accepted format. (Of course, almost any biometric template can be embedded in the document, but the advantage of dsv technology is its similarity to the traditional method, where a person signs their name, rather than gives their iris scan.)

Dsv technology can also be used as a possible password replacement tool for accessing computers, and increasingly it is being used to access electronic devices, such as PDAs, where a screen and stylus already exist.

Biometric technologies looking at the signature use both the static features of the signature shape and the dynamic features of pen and hand movements. There are hundreds of static features that are important for a reliable verification process, such as inclination, line density, line vectors, compactness and enclosed areas. In fact, there are up to 50 primary and 600 secondary parameters that can be extracted, calculated and weighed in a parameter vector that characterises an individual's identity.

Pure dynamic signature analysis can analyse factors such as shape, speed, pressure, timing and stroke pattern to establish the unique patterns that characterises an individual's identity. With dynamic analysis, taking into account timing development of pen position and pressure, it becomes very difficult to reproduce the natural movement of the true signer, especially if they only have a copy of the signature as a model. (Greater forgery success has been noted when a person has been able to study the person in the act of signing.)

There are numerous companies in the dsv market spread across the world. Some are focusing on dynamic analysis alone, while others combine their authentication decisions on both static and dynamic characteristics. Unlike the technology, the market, to date, has not been dynamic, but the potential for greater success is clearly present, as long as it can overcome certain technological and market barriers.

3.3.5.1 Technical Developments

Dsv technology is ideal in certain circumstances, but wholly unsuitable in others. Dsv is not recommended in situations where identification rather than verification is required, for example. But where a person is already expected to sign their name – when filling out a form, or signing a contract – it can be an excellent verification choice. It not only enhances the security of the act of signing, providing good non-repudiation properties, but it also allows that document to be stored electronically rather than taking up valuable space in a filing system.

Input devices

One of the biggest technological developments in the dsv industry is in the input devices that are used to capture a person's signature. Traditionally, most input devices have been tablet based – where a pen (proprietary or not, depending upon the tablet) is used to sign on a special pad that can pick up different characteristics, such as pressure. Other options exist, however, such as pressure sensitive TFT-displays or non-pressure sensitive handheld devices such as Palms or Pocket PCs. It is also possible for the pressure and input to be recorded inside a special pen that also monitors the three dimensional space near the paper, but there has been limited commercial success with this option to date.

The latter half of 2002 will see the introduction of Tablet PCs from Microsoft, which will enable users to write on its screen like traditional pen and paper. In late June 2002, Fujitsu, Motion Computing and Toshiba demonstrated prototypes of the Tablet PC. It is likely there will be a substantial increase in the availability of these capturing devices in the future and they could provide an exciting new platform for dsv software companies to work with. Various smart phone producers are also looking at implementing capturing technology into their next generation of products.

The quality of data captured from these input devices varies widely and as of yet there is no standardisation for these types of device. It is quite likely that a larger institution will use various input devices like a mixture of tablets and PDAs. But as PDAs on the market in early 2002 did not support the capturing of different pressure levels and registered less signals per second than a quality pen pad, verification software must classify the input signals given from any input device.

The price trends for tablets – with or without a display – are continually moving downwards. For example, at the high end of the market a tablet with display from a leading supplier like Wacom or Interlink, might cost around US$2000. In 2000, a typical price was almost double at around US$4000. At the lower end of the price spectrum, some tablets are now sold for less than US$25 in discount chains, however they can lack robustness and have other limitations.

In looking at tablets there are certain quality criteria, depending upon the application, which are useful to consider, such as:

- good similarity between the display image and the signature on paper;
- a sufficient number of captured time signals throughout the signing process;
- a differentiation between various writing pressure levels;
- the secure and unambiguous communication between the capturing device and the processing unit.

Inconsistent behaviour

Signature verification, like voice recognition, is a behavioural biometric and depends on the consistent behaviour of the individual for the greatest success. Inevitably, however, there is a degree of variability each time a signature sample is obtained. For some individuals this variability may be very high and the system may be unable to cope.

Various external influences are able to cause a change in this typical behaviour. If an input device can deliver an exact optical feedback as the person signs their name, this is an advantage. If the user realises that their signature varies from the usual result on paper then they often will alter their way of writing, which is undesirable.

When using a pad/tablet without a display, first-time users may experience difficulties when writing on a clean surface, even though the captured signature is visible on a monitor nearby. To avoid this problem some suppliers are now offering the option to use paper on the pad and write with a normal pen or a special inking pen.

Some suppliers have been looking at assessing the quality of the signature models constructed during enrolment. During enrolment, a user is normally expected to submit a number of samples. By analysing the pattern of feature distribution it may be possible to determine whether the user is exhibiting consistent behaviour. This would possibly allow for the detection of 'goats' (people who's signatures are hard to classify and may contribute to high system error rates) and the design of systems to prompt the user for more samples. Some suppliers have installed systems that are able to update templates on an ongoing basis – adding to the template model each time the user gives a good signature (this also helps with template ageing issues as the template is updated over time).

3.3.5.2 Market Developments

Signature based biometric technology is making strong headway in two main horizontal markets – transactional-based applications and PC/Network Access (including access to personal digital assistants).

The vertical market where dsv systems are proving most popular is the financial services sector, including insurance companies and banks. Notable contract wins over the last year have been the implementation of over-the-counter dsv systems in banks. These include the well-publicised Bank Hapoalim installation in Israel (provided by WonderNet), with an estimated 5,000 seats. Another similar installation is still secret, but will involve a major financial services institution in the UK.

In this particular market, a typical set-up is to position signature tablets at the bank tellers' counters. Enrolment into the system typically requires the account holder to provide their signature a number of times in order that a single profile can be generated and stored in a central database. Subsequently, when a user comes to the bank, the teller enters their account details, which are then retrieved along with their template. The customer is then asked to sign to allow a real-time comparison to be made.

Companies such as Softpro in Germany look at both the static and dynamic properties of a person's signature and use these properties to verify signatures from cheques in an offline environment to identify possible fraud.

Another major market force that will promote the use of dsv systems in the future is the desire to improve corporate data security coupled with the need for reducing operating expenses. One way to do this is to produce and process documents electronically. This not only speeds up the process, but also reduces the costs of paper handling.

Although the move towards a paperless environment may be taking longer than expected, the need for verifiable signatures has become imperative if the documents are to be legally viable. Cutting and pasting of a person's scanned in signature is clearly no good, as anyone can do this if they get a copy of that person's signature. With a signature verification system, information about the signature is stored encrypted together with the document. There is often the possibility to integrate this information with various forms of electronic signatures. If the content of the document is changed in any way, this will automatically be indicated.

According to one industry statistic some 30 billion documents arise in the USA alone, so the costs associated with their moving and storage is significant and the market opportunity for dsv technology is clear. By way of an example, a large insurance company using this sort of technology has reportedly processed over one million forms and saved US$30 million in the last two years. The forms, which averaged three or more signatures, were said to have reduced in cost from approximately US$32 per form to less than US$2.

Events such as September 11th and the subsequent anthrax scare left countless paper records lost or destroyed. This generated a greater incentive for a more paperless environment in the short term, but in the medium term it is the cost savings argument, not one-off disasters, which will likely push the market forward. That said, according to some industry suppliers, such as CIC, there is now more interest in looking at these cost implications since September 11th.

Other applications for the technology are numerous, but include acknowledgement that services, such as a parcel delivery, have been received; the access of controlled documents; logging into electronic organisers; and credit card fraud reduction through integration of dsv into payment authentication software.

A selection of the leading market players are:

- Communication Intelligence Corporation (CIC)
- Softpro
- WonderNet
- Cyber-Sign

3.3.5.3 Drivers

- *Legal recognition of electronic signatures* – Legislation now exists in most countries, which makes electronic signatures legally binding. DSV technology is only one form of acceptable electronic signature, but its similarity to pen and ink signatures makes it a good choice for those users wanting to make the process as natural as possible.
- *Reference implementations growing* – The number of high-profile reference implementations using signature verification technology is growing and looks set to increase in the future. This gives vital credibility to the technology, which has had its doubters in the past. The banking and insurance industries, in particular, look set to be a major adopter of this type of technology in over-the-counter applications and in helping reduce paperwork for insurance agents in the field.
- *Good potential cost savings* – As seen in the above text, the potential for cost savings in moving towards a paperless environment can be significant. In a

paperless environment, the use of dsv characteristics is desirable as it enables the document to have a verifiable and legally accepted signature, which becomes invalid if the document changes in any way. The cost savings through a reduction in fraud are also a driver for financial institutions. A potential future driver may be the reduction of internet payment fraud.

- *Well accepted biometric by users* – People have signed their names in order to make an agreement or transaction for hundreds of years. The association between normal signatures and electronic dsv signature makes it highly acceptable to users.
- *Resistant to attack by forgers* – A dsv-based signature template contains a large amount of information, which helps the system make an accurate decision as to the legitimacy of a fresh sample. Even an average signature is almost impossible to forge by a mimic, although a simple signature could be forged if the forger has studied the way a person tends to sign. Nevertheless False Acceptance Rates are generally low.
- *Increasing number and lower cost of capture devices* – As seen above the number of devices able to capture dynamic signature properties is proliferating and costs of high-quality devices (best for any serious financial implementation) are reducing.

3.3.5.4 Detractors

- *Poor performance due to signature variability* – The environment in which a person signs their name can impact the quality of a person's signature, such as whether they are standing up, sitting down or if they are in a hurry. Some people have a tendency to be erratic signers regardless. This is a major consideration, because there are a number of choices – reject the user and face the consequences if they are indeed the right person, or reduce the tolerances for an erratic signer, which reduces the security of the system.
- *Competition with other digital signature technologies* – Other digital signature technologies, such as PKI systems or other types of biometrics, are also common and pose a significant viable alternative for companies wishing to move to a paperless system.
- *Unable to penetrate traditional biometric market strongholds* – The biggest horizontal market to date for biometrics has been physical access control and the strongly developing vertical markets are based around the government-led ID schemes and travel industry. Dsv systems are not generally well suited to these types of application and revenue generation could suffer proportionately compared with other technologies as a result.
- *Slow move toward the paperless office* – The paperless office is a major opportunity for dsv technology, but it has been slow to emerge. It is a major and expensive undertaking for a company to shift its systems to a paperless system. While the cost benefits are clear, the initial outlay in capital and time/effort is a problem. In many instances, such a move requires a major commitment from the management team. Until recently there have been more compelling IT issues to face, such as the Y2K bug and preparing for the introduction of the Euro.

3.3.5.5 Outlook

Signature verification provides a natural, secure and cost effective way for companies to enhance security in many ways. From cheque processing and customer verification at financial institutions to eliminating paper contracts and documents in the office, the technology is set to grow strongly over the next few years.

In the short term the use of dsv in financial environments and on PDA devices is expected to be the main growth area. However, as the move towards removing signature-based paper documents continues in office environments, this could prove to be the exponential growth driver the industry is seeking.

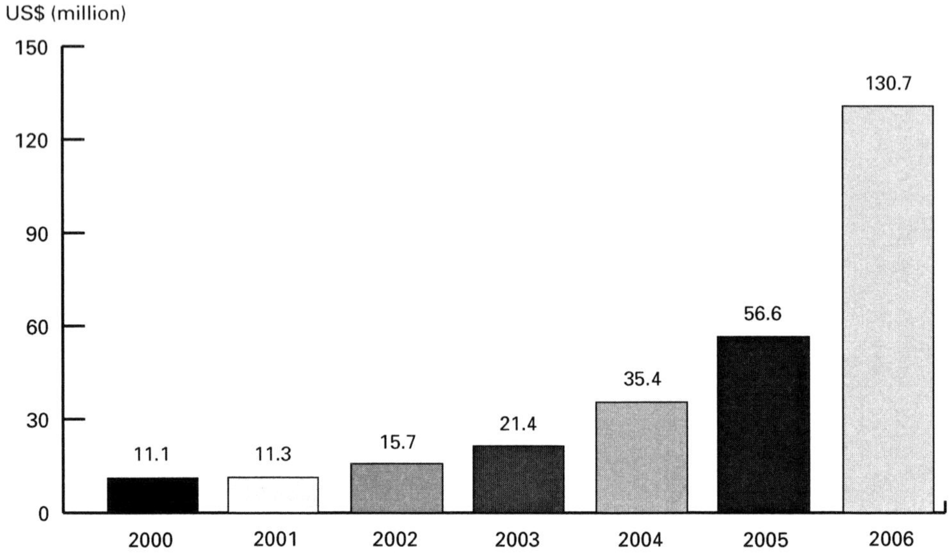

Figure 3.8 Revenue Forecast for Dynamic Signature Verification Technology 2000–2006 (US$ million)

In revenue terms the dsv market will move from US$15.7 million in 2002 to US$130.7 million in 2006, by which time it will represent some 7% of the market.

3.3.6 Iris Recognition Market

Iris recognition is a fast, accurate, secure and potentially cost effective biometric technique that can operate in both verification and identification modes. It is being used to recognise users in a wide range of market sectors, but despite technological and market developments over the last few years, the market has not yet taken off in a spectacular fashion. Nonetheless steady progress is being made and this is expected to continue for the foreseeable future.

The iris is an excellent source of identity information. The pattern is both stable and unique to an individual and is not normally prone to damage or wear – unlike some other biometrics, such as fingerprints. Iris recognition devices use a video image of the eye, which is typically captured by a standard video camera. Once captured, iris recognition matching algorithms examine more than 240 degrees of freedom in the human iris to create a 512-byte data template, normally known as an IrisCode record.

One main supplier, Iridian Technologies, dominates the market for iris recognition technology. This US-based company holds worldwide patents for the processes behind iris recognition, including the matching algorithms. Despite this dominance, however, the market has many other players that have developed business relationships with Iridian Technologies. These include hardware manufacturers, software partners, vertical market partners and value added resellers. Increasingly Iridian Technologies is focusing on software provision (see Iridian in *Company Profiles*).

Iridian's main relationships include:

Hardware
Panasonic
LG Electronics
OKI

Software
Bioconx
BioNetrix
Computer Associates
IdealSoft
I/O Software
Keyware
Netegrity
Saflink

Vertical markets
Government/Public safety/Justice – Tier, Radian
Transportation/Immigration – EyeTicket Corporation, Joh. Enschede, Unisys, Argus Solutions, BioID Technologies, NEC
Workforce management – Datamatics

VARs
SAIC
Software Spectrum
TigerDirect.com
Politec
Lockheed Martin

3.3.6.1 Technical Developments

As a relatively new commercial entry to the biometric market, there are still plenty of technical developments being made in the iris recognition sector.

Iris recognition systems can operate in either an automatic capture mode or they can be used to capture the iris image manually. The automatic capture systems are easier to use (although more expensive), as there is no need for the user to assist with the aligning and focusing of the camera. Clearly a manual system is more difficult, as the user must move the sensor (or themselves) back and forth in order for their iris image to be captured.

Designing better image capture systems is a trend that is evident at present. One of the most recent announcements was by OKI Electric Industry (in association with Matsushita Communication Industrial Co). It launched a system (the *Irispass-WG*) which can automatically detect the position of a person's eyes and then scan the iris and the pupil. Panasonic is another supplier to have launched a device – the *BM-ET500 Series* access control system – which essentially operates in the same way by capturing the iris image in an automatic, user-friendly way.

Another market trend has been the launch of small, less expensive iris recognition devices. The best-known device is the *Authenticam* from Panasonic. The small size of the device and its compatibility with a range of middleware products, such as I/O Software's *SecureSuite*, makes it ideally suited for computer security applications. It can also be used as a Web camera. Another small IT device is the *Irispass-h* from OKI Electric Industry.

As a small note of caution, spoofing attempts made on the Authenticam device proved successful in the first half of 2002, although the manufacturer is believed to have made moves to rectify the situation by adding more of the anti-spoofing measures found in more expensive systems.

Away from the hardware developments, work in the software field has been continuing. Of particular note is the development of a new technique for altering the format of an iris template. This may prove important in alleviating some of the privacy concerns surrounding the use of large biometric template databases – in particular function creep. The idea is that a template's format can be made specific to a particular application, making it unworkable in other applications. The process of changing the format does not change the efficiency of the matching process, so no degradation to performance occurs.

3.3.6.2 Market Developments

Iris recognition technology is making headway into numerous horizontal markets including physical access control, computer security, citizen ID and transactional-based applications. Border control, simplified passenger travel and access control applications at airports are today the most talked about use of the technology, in particular since the terrorist attacks of September 11th.

The main markets for iris recognition devices are:

- Airport applications, including verification of passenger identity for fast track passenger processing and access control for staff;
- Securing computer systems in places such as hospitals and government offices;
- Physical access control in numerous locations, such as prisons, bank vaults, government buildings etc;
- Financial services industry applications.

Although Iridian Technologies is the dominant player in the iris recognition market, primarily thanks to its strong patents (US Patent 4,641,349 entitled "Iris Recognition System"), competitors have emerged that are hoping to capture significant market share. For example, Senex Technologies from South Korea, has launched a device called *TrueEYE*, which uses alternative iris acquisition and processing algorithms to Iridian's technology. Iridian claims this company is contravening its patents and was taking action at the time of writing this report.

Senex is a public company and is targeting the ATM market, online user identification market and access control-based applications. Its technology only extracts 128 bytes of information and is still in the final development stage.

3.3.6.3 Drivers

- *High accuracy* – Iris recognition is the most accurate of biometric technologies. This was borne out by tests performed by the Centre for Mathematics and Scientific Computing at the UK's National Physical Laboratory (NPL), where the technology did not falsely accept any users. This is a strong driver, especially in applications dealing with large populations or where security is of the highest priority. The accuracy of the device allows the technology to work excellently in identification mode, which allows the user to use the device hands-free.
- *The increasing number of distribution channels* – As seen above, there are now many outlets for iris recognition technology. This is a major driver, as the combined and often targeted marketing efforts by the various VARs, software/hardware partners and vertical market solutions providers will serve to raise awareness and lead to enhanced sales activity.
- *Raised interest from immigration/travel industry sector* – Since September 11th, 2001, there has been a vigorous interest in the potential use of biometrics in the immigration/travel sector. Installations to enhance simplify passenger travel at the UK's Heathrow Airport and at Schipol Airport in the Netherlands have all been well received. The uniqueness of the iris means it works extremely well with large populations of users. Therefore the use of iris recognition technology in passports is a distinct possibility.
- *Launch of less expensive devices* – In the past iris recognition devices were all very expensive. Now less expensive devices exist in the IT security field, such as Panasonic's Authenticam device. Although it is unlikely to make a dent in the consumer market, organisations are being persuaded to use the devices to secure confidential computer-based information or controlling access to intranets etc.
- *Hardware with automatic iris-finding algorithms* – Although a bit more expensive, the launch of easy-to-use systems that don't require user intervention will be attractive to some customers, especially where users will be from non-technical backgrounds.

3.3.6.4 Detractors

- *Price of units* – Although there are IT security devices on the market, such as the Authenticam, which retails as cheaply as US$170 (but more typically at US$350), the price of a typical iris recognition-based access control system is still significant. Depending upon the system, a standalone unit with automatic iris location capability would retail at around US$10,000, while a multi-door system would be in the region of US$7000 for each extra door. Iris-based door access systems with less high-tech electronics (from LG, for example) are available for less. Regardless of the system, the price is still a big initial outlay compared with all other access control systems, such as swipe/smart cards, PIN pads etc. This means that customers have to be fully convinced as to the benefits of the biometric system in terms of long-term payback and enhanced security before committing.

- *User perception still a problem* – Although user perception of iris recognition devices has improved, the technology is still seen as much more of a science fiction-based technology than other biometrics, such as fingerprint or facial recognition. This means that iris recognition may not even be on the radar of some potential purchasers or that they may see the technology as overkill. There is also a general misunderstanding on how iris recognition devices operate (i.e. some people still think a laser light scans the eye, which is untrue). Education is still vital in this market.
- *Monopoly on the matching algorithm side of the business* – The fact that Iridian Technologies owns all the iris-recognition-based matching algorithms (except for companies such as Senex Technologies, see above) is a cause of concern. Although it is likely that suitable reassurances are in place, some large organisations or government agencies may not like the idea of being reliant on a single company for an important building block in their potential system.
- *Spoofing of Authenticam device* – Although this spoof was achieved after a concerted effort and on this relatively 'low-tech' device (by iris recognition product standards), it was still a blow for the iris recognition industry. Previously quoted "zero false acceptances" can no longer be claimed. Although the manufacturer is believed to have altered the technology to prevent the particular spoof technique from being repeated, it could still be a sticking point for potential purchasers.
- *Interaction required and good conditions needed* – With most iris recognition products, the technology does need a level of user co-operation with the user holding still for a moment. The technology is also constrained by the requirement for decent environmental conditions, such as adequate lighting etc.

3.3.6.5 Outlook

The industry's most accurate biometric, iris recognition has made a successful launch into the mainstream and this surge looks set to continue. Prices have come down, and the number of hardware vendors has proliferated, extending the reach of the technology in numerous vertical market sectors.

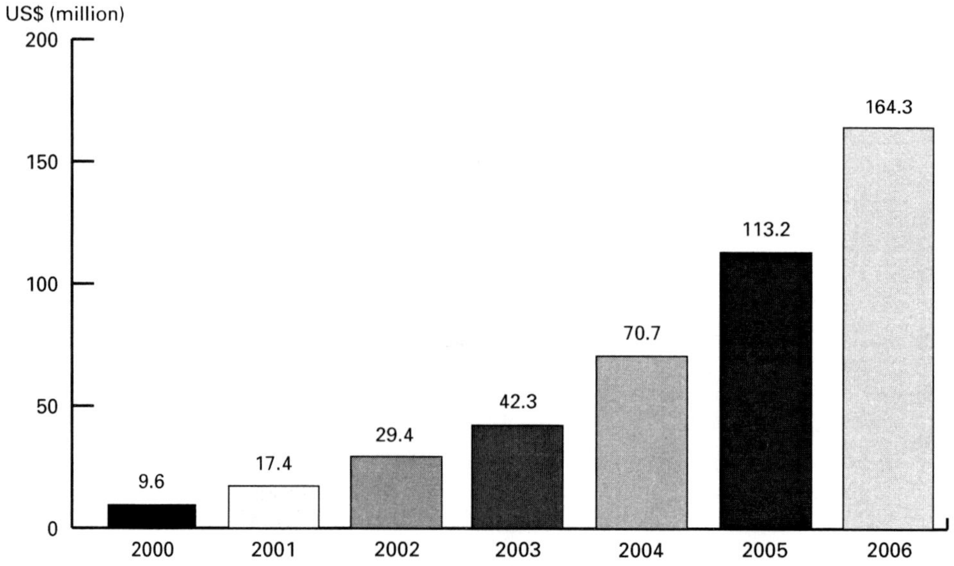

Figure 3.9 Revenue Forecast for Iris Recognition Technology 2000–2006 (US$ million)

The ability of the technology to distinguish individuals in identification and verification mode, without a false match puts the technology in a strong position for widescale high security biometric implementations, which will likely emerge over the next three years in the wake of September 11th.

By 2006 revenue will have reached US$164.3 million from a base of US$29.4 million in 2002. This will represent almost 9% of the biometric market.

3.4 Horizontal Market Forecasts

This edition of the Biometric Industry Report has segmented the industry into technology, horizontal, vertical and geographic markets. The horizontal markets show 'how' the technology is being used in the numerous applications it is now employed.

The different horizontal markets have been defined as:

- Physical access control/time attendance
- PC/network access
- Financial transaction authentication
- Citizen identification
- Surveillance

The first two categories are self-explanatory as is the surveillance category. The financial transaction authentication sector includes e- and m-commerce, ATM transactions, point of sale transactions and so on, while citizen ID includes the identification of citizens, immigrants, asylum seekers, employees etc.

Any manufacturer submitting figures not covered by these categories had the choice of an 'other' category.

3.4.1 Outlook

Figure 3.10 shows that the two main horizontal markets are physical access control/time and attendance (PAC) and PC/network security, and that these will remain the major horizontal markets over the reporting period. Interestingly it can be seen that the PC horizontal was actually larger than the access control horizontal in 2000, but that access control regains supremacy in 2001 and 2002, only to be permanently overtaken in 2003.

The PC/Network access market had been moving strongly in 1999 and 2000, nudging it ahead of the PAC horizontal in 2000 for the first time in the industry's history. The reason for the PAC horizontal taking over for the next two years can be explained more as a fall off in demand for IT security goods than a spurt in growth in access control installations.

In fact, access control installations will continue to exhibit steady – but not spectacular – growth over the forecasting period. PC/network access installations,

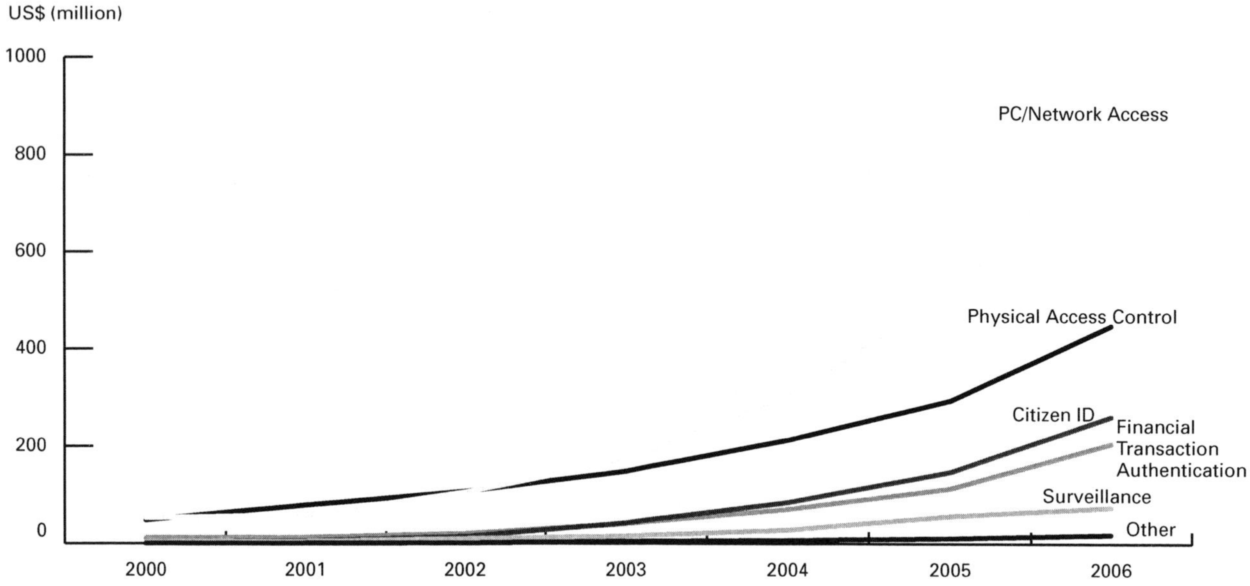

	2000	2001	2002	2003	2004	2005	2006
Physical Access Control	48.2	78.6	107.2	148.5	212.2	294.2	448.1
PC/Network Access	52.6	58.5	105.5	170.7	304.1	509.2	858.9
Financial Transaction Authentication	11.8	12.8	20.8	41.0	70.7	113.2	205.4
Citizen ID	3.2	5.5	14.7	42.7	84.9	147.1	261.4
Surveillance	3.7	4.4	10.3	15.7	28.3	56.6	74.7
Other	0.4	0.6	1.5	4.9	7.1	11.3	18.7

Figure 3.10 Revenue Forecast for the Horizontal Markets 2000–2006 (US$ million)

however, are set to grow strongly from 2003, reaching clear ascendancy by 2004 and beyond.

The other horizontals will also become more predominant by 2006. In particular the financial transaction authentication and citizen ID horizontals will create significant revenue for the industry. Looking forward even further, it is likely that the financial transaction horizontal will eventually become one of the top horizontals, as the use of biometrics gets increasingly used for e- and m-commerce, although this will take longer to establish than 2006, however. Citizen ID applications are a pressing issue today and so the growth in this horizontal is expected to be rapid from this point onwards.

Surveillance applications have grabbed many of the headlines over the last year, and the horizontal is experiencing strong growth. However, the total number of potential installations available to vendors limits the size of the market. So as a proportion of the industry, this is likely to remain one of the smaller horizontals.

3.5 Vertical Market Assessments and Forecasts

The underlying dynamics of a market can be found by assessing the vertical markets. Looking at all the potential drivers and detractors gives strong indications as to where the fast growing sectors will be in the future.

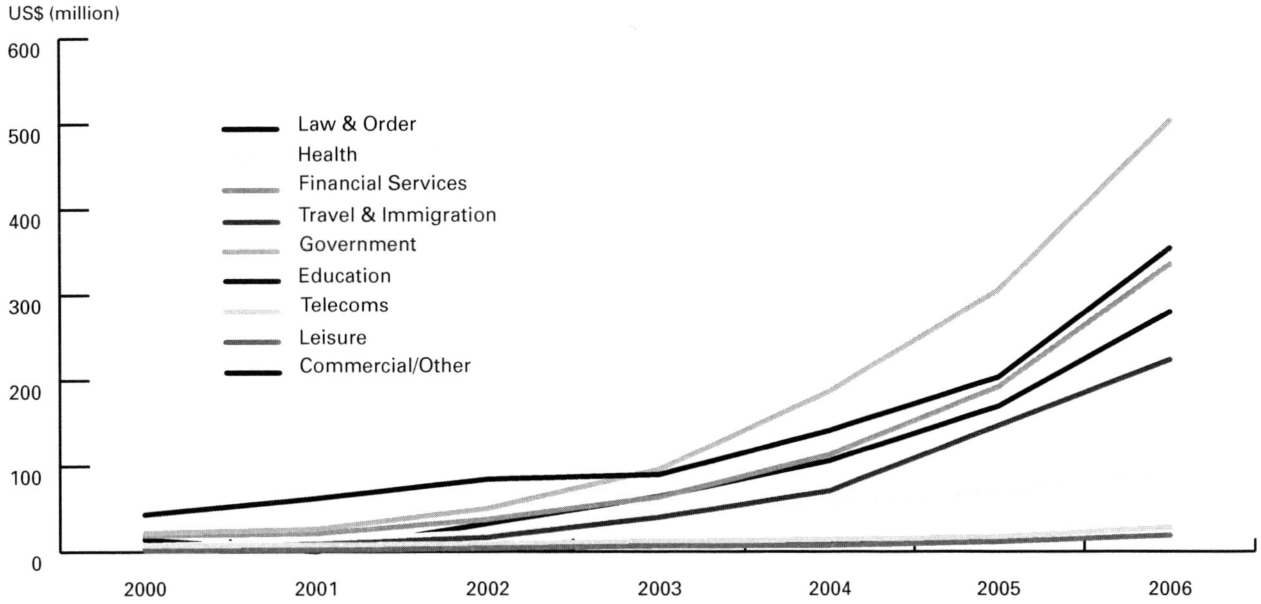

	2000	2001	2002	2003	2004	2005	2006
Law & Order	14.1	18.2	32.7	64.4	106.1	169.7	280.1
Health	7.4	12.4	23.1	44.7	56.6	67.9	93.4
Financial Services	18.9	21.1	37.5	63.5	113.2	192.4	336.1
Travel & Immigration	4.7	8.7	16.5	39.9	70.7	147.1	224.1
Government	21.7	26.2	50.2	96.0	187.4	305.5	504.1
Education	2.2	2.1	2.7	7.1	10.6	17.0	28.0
Telecoms	6.6	8.0	8.9	11.8	14.1	17.0	28.0
Leisure	1.3	1.8	4.1	6.5	7.1	11.3	18.7
Commercial/Other	43.1	62.0	84.4	89.8	141.5	203.7	354.8

Figure 3.11 Revenue Forecast by Vertical Markets 2000–2006 (US$ million)

While the horizontal forecasts look at how the biometrics will be used i.e. physical access control, a vertical market focus looks at where the biometrics are being used – physical access control in a hospital, for example.

As can be seen in Figure 3.11, the strongest market between 2002 and 2006 will be the government market. Already, the promise of significant revenues exists, especially in the USA – although the lead times for this type of work (especially in the larger nationwide schemes) are notoriously long.

Unsurprisingly the travel and immigration sector looks set to do better than earlier anticipated. In the aftermath of September 11th, many countries have accelerated their planned adoption of high-tech schemes, whether for simplifying passenger travel, or for producing more secure passports. Surveillance in areas such as airports is also likely to accelerate in the next few years, as the technology in existing implementations gets a fair trial period, allowing for the technology to be tweaked to a performance level acceptable in such environments.

Financial services and law and order applications will also show good growth, as will the commercial market, which incorporates those applications not covered by the individual categories.

Other markets, such as healthcare, will not be dominant in the future, but are still generating spectacular growth potential in the near term. This is most likely thanks to the impending HIPAA regulations, which come into force over the next 12 months.

The leisure, education and telecoms markets are not expected to show great potential in the short term, although the telecoms market clearly has a great deal of latent potential as the introduction of new 3G-based phones comes in.

The following sections will discuss the main application areas and the main drivers and detractors that are likely to influence the take up of biometrics over the next four years.

3.5.1 Education

The education market provides some interesting long-term opportunities for the biometrics industry. Looking at all vertical markets for biometrics, one of the key detractors is the lack of knowledge of the technology by members of the general public. If biometrics can develop awareness in the educational sector, this could – in the long term – reap rewards for the industry as a whole.

The growing use of IT throughout the education sector means that many institutions already have the basic infrastructure from which to develop biometrics. This has resulted in pockets of interest in the use of low-cost devices (especially fingerprint technology), in schools and colleges in the US and Europe. These devices are gaining ground in low security areas such as libraries and canteen facilities. However, most schools are acutely cost-conscious, so if biometrics are to make further inroads in these low security areas, costs still need to fall further. In addition to these areas, suppliers are also witnessing a greater degree of interest in access control devices as fears for student safety grow.

3.5.1.1 Summary of Market Applications

The following is a list of typical biometric applications in the education sector:

- Access control to facilities
- Prevention of cheating
- Replacing school IDs
- Replacing lunch passes
- Replacing library cards
- Enhancing attendance rosters
- Supplementing time cards
- Computer access control for students and staff
- Detect whether employees have criminal records (using livescan equipment to acquire fingerprint – not considered in this report)
- Online learning

3.5.1.2 Drivers

- *Growing technology use* – A strong driver for growth within the education market appears to be the growing use of technology. This affects all aspects of education: Attendance records and general administration are frequently maintained on a PC; learning centres need to limit access to sensitive parts of their networks as well as the internet; and a growing demand for online learning within both the public and private sector has spurred on the need to ensure that a student taking part in a programme of study is actually the person who enrolled.

- *Cost efficiency* – Technology that can help drive down costs is needed within the educational world. As the cost of biometric devices falls, they will combine efficiency and ease of use in a number of areas, such as the replacement of time cards and overcoming password delays.
- *Safety and security concerns* – Security and access control are fairly strong drivers, especially in universities. Since September 11th, safety concerns have risen along with the desire to protect access to sensitive laboratories. A major advantage of biometrics over other forms of access control – such as tokens or PINs – is that employees or students who have resigned or been dismissed, can instantly have their template removed from a system and access barred.
- *Removal of social stigma and risk of theft* – In areas such as school canteens, biometrics are seen as positive by parents and children alike. From the parents' point of view, biometrics can eliminate stolen or lost lunch money and lunch cards. In some cases they can help parents to monitor and restrict what their children are eating. They also reduce the stigma of free lunch programmes by removing the need for special vouchers or cards. As prices fall, this could become a moderate driver for the industry.

3.5.1.3 Detractors

- *Cost* – The biggest detractor within this market is cost – especially for organisations operating in the state sector.
- *Privacy issues* – Privacy concerns continue to be a fairly strong detractor. However, in the US, the Family Educational Rights and Privacy Act protects students from the sharing of data with law enforcement and government agencies. Since biometric data is considered a school record, this 1974 privacy law requires a student's parent to sign an agreement for the data to be released, safeguarding the possibility of data being passed to any third party.
- *Dislike of fingerprinting* – Although a fairly weak detractor, the negative association that some people have between fingerprints and criminality could deter system acceptance.

3.5.1.4 Market Outlook

The education market will continue to provide attractive opportunities, especially in the control of access to both resources and educational establishments themselves. Costs still have to be reduced further, and decision-makers need to be made aware of the technology, but in the medium to long term long term, it is anticipated that growth will accelerate and help the industry as a whole in the long term. Worldwide revenue in 2000 stood at US$2.2 million for this sector, but this is anticipated to reach US$28.0 million by 2006.

3.5.1.5 A Selection of Education Market Installations

Table 3.3 A Selection of Education Market Installations

Name	Technology	Name of supplier	Status	No. enrolled
Sweden				
Stockholm City	Fingerprint	Saflink	Pilot	460
Name of system: SAFmodule for NMAS **Function:** PC user authentication.				
Switzerland				
Morges	Fingerprint	–	–	–
Function: Purchasing food from school canteen.				
UK				
Royal Grammar School, Guilford	Fingerprint	Identix	–	–
Name of system: Fingerprint technology from Identix. **Function:** Access control.				
USA				
Various Chicago schools	Fingerprint	Identix	Pilot	–
Name of system: FingerScan V20 from Identix. **Function:** Testing the use of biometrics for physical access control.				
Continueded.com	Keystroke dynamics	NetNanny	–	40,000
Name of system: Biopassword from NetNanny integrated with ContinuedEd.com's `TicketErasers' technology. **Function:** Course enrolment.				
Eagen High School, Minn	Fingerprint	–	–	–
Function: Check out library books.				
Florida	Fingerprint	Lockheed Martin Information Systems	–	–
Name of system: Digital fingerprinting equipment from Lockheed Martin Information Systems. **Function:** Detecting employees with criminal records.				
Various Pennsylvania schools	Fingerprint	Sagem	–	–
Name of system: System deployed by Food Service Solutions using scanners from Sagem. **Function:** Several schools within Pennsylvania have adopted fingerprint technology to pay for school lunches.				
Triana Science and Operations Center, University of California	Fingerprint	Bioscrypt	Live	–
Name of system: V-Pass fingerprint units from Bioscrypt. **Function:** Protect entry to sensitive facilities, which have direct links to NASA computers.				

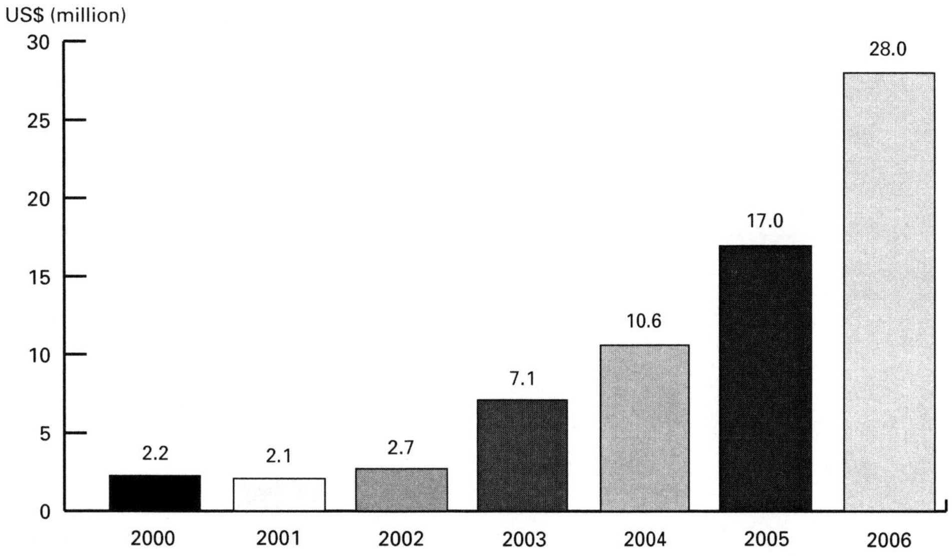

Figure 3.12 Revenue Forecast for the Education Market 2000–2006 (US$ million)

3.5.2 Financial Services

The financial services sector has traditionally provided the biometrics industry with the opportunity to provide some of its most imaginative solutions to security and convenience problems. News of iris recognition at ATMs, for example, created media headlines and helped expose the general public to the idea of biometric technology in general. Pockets of interest have been seen in areas such as check cashing and over-the-counter solutions. In spite of this, however, it is an area of business that has often failed to live up to media hype. Most biometric work continues to take place behind the scenes in areas such as staff access to computers, vaults and processing centres.

3.5.2.1 Summary of Market Applications

The market as a whole provides two potential areas of business: Back office functions and customer-facing applications. These two areas are vastly different and require completely different approaches from biometrics vendors and financial institutions alike.

Back office applications include:

- Computer/network security
- Non repudiation
- Physical access control to vaults, processing centres, teller environments and safe deposit boxes
- Insurance companies – reducing paperwork

Customer-facing applications can be listed as:

- Point of Sale (POS) devices
- Electronic purses
- ATM machines

- Authentication at the bank teller
- Telephone banking and other telephone transactions
- Internet banking and other internet transactions
- Check cashing
- Supermarkets with self-cashiering checkout lanes

3.5.2.2 Drivers

Currently, the main drivers exist in back office and staff functions. This is because most banks find it easier to implement systems within a staff environment than in a customer-facing area where opinions and expectations have to be carefully considered.

- *Staff physical access control* – The need to control staff access to sensitive areas has been an area of growth for several years and is expected to provide a substantial amount of revenue in the next few years.
- *Non repudiation* – Providing a method of non-repudiation allows each financial transaction to be accurately attributed to an individual member of staff and customer. This allows fraud to be reduced, money saved and insurance premiums lowered. Insurance companies are showing themselves keen adopters of the technology within this area.
- *Customer feedback* – Customer opinion ultimately drives the use of biometrics in ATMs, call centres, POS devices, check cashing, over the counter solutions and internet transactions. As customers become exposed to biometrics in other areas of their lives (such as automobile security, computer access etc), they are more likely to be positive about their use within the financial service environment. Early indications show positive feedback, especially since September 11th. As a result, more financial institutions are beginning to consider the use of biometrics. However, this opinion may only be temporary.
- *Remote transaction authorisation* – The increased use of remote and automated transactions could help propel the demand for biometrics in the financial services arena in the long term. For now more mature technologies are still preferred by most parts of the financial services industry, but the increasingly changing nature of banking may help the biometrics industry to find a niche. For example, biometrics could provide convenience and help to reduce the amount of bank card fraud committed through remote transactions.

3.5.2.3 Detractors

- *Cautious about technology change* – The cautious nature of many financial services organisations continues to provide the strongest barrier to growth. This caution is matched by an on-going commitment to other forms of authentication such as PINs and chip cards. In particular, many banks have committed to complex and expensive security solutions within the last 3–5 years. Most financial services companies will be unwilling to invest heavily in biometrics until they have seen how other organisations have received the technology.
- *End user rejection* – A potential lack of end-user acceptance is also a strong detractor for banks considering the roll-out of biometrics in customer areas. In particular, most banks have recognised that false rejection can be very

damaging in a customer-centric environment. Until the industry is convinced that end-users will have a positive experience of biometrics through the use of robust and accurate biometric technology, it remains reluctant to invest in the technology.

- *Template management issues* – The financial services sector is concerned with the perceived technical difficulties surrounding the capturing and management of large numbers of biometric templates. This includes the fear of identity theft and the question of privacy. These issues are of particular concern for customer-facing operations.
- *Business case* – The biometrics industry often lacks a solid business case for developing a broad-based implementation of biometrics. This business case needs to identify the appropriate technology to suit the application. At the same time, the biometric being used needs to be convenient, highly accurate and cost-effective.

3.5.2.4 Market Outlook

In the short term, it is likely that employee access to networks and other facilities will offer the greatest market for biometrics in the financial services sector. Call centres using a combination of speech recognition and biometric speaker verification may also provide a popular way of introducing biometrics to customers.

The further use of biometrics within customer-facing applications will continue to emerge slowly, although they are unlikely to supply the opportunity for widespread growth in the medium term. This is because the financial services sector will be guided by other industries and will not risk losing customers until it is confident of the popularity of biometric technology. In particular, ATMs are unlikely to be a large area of business for a considerable amount of time due to the cost involved in changing these systems and the lack of public enthusiasm for such a change.

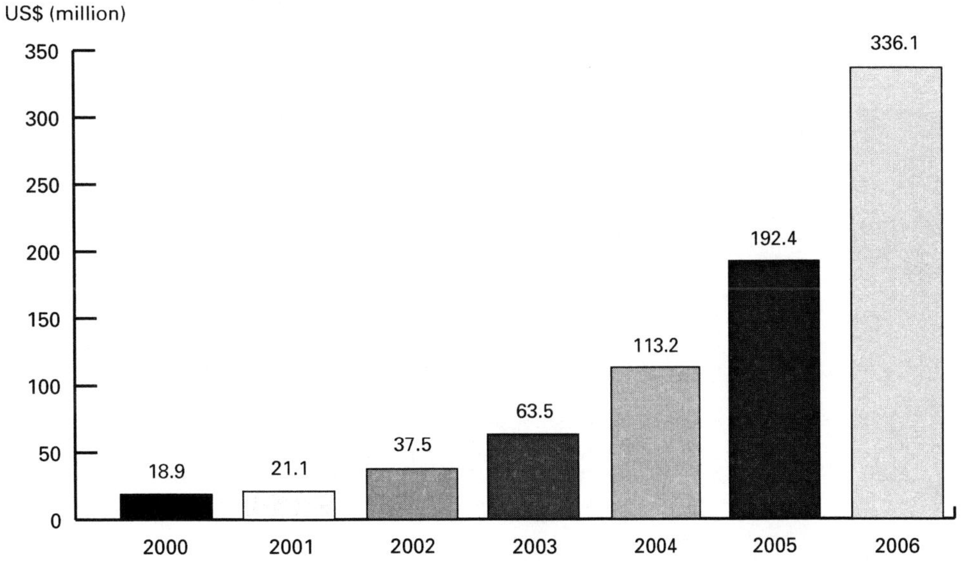

Figure 3.13 Revenue Forecast for the Financial Services Market 2000–2006 (US$ million)

The large number of applications in this sector make it significant in terms of revenue and overall market share, which stands at around 15% of the biometric industry. From a base of almost US$19 million in 2000, growth will be steady over the next few years, accelerating into 2006 when revenue will hit an estimated US$336 million.

3.5.2.5 A Selection of Financial Services Market Installations

Table 3.4 A Selection of Financial Services Market Installations

Name	Technology	Name of supplier	Status	No. enrolled
Brazil				
Brazilian Mercantile & Futures Exchanges (BM&F)	Fingerprint	Identix	Live	–
Name of system: Fingerprint security system.				
Function: Secure access to electronic Global Trading System.				
Banco Bradesco	Speaker	Nuance	–	–
Name of system: Nuance Verifier from Nuance used in Fone Facil service. The system has been designed by Nuance and its Brazilian partner GMK Eletronica.				
Function: Increase security of phone-based banking services.				
Chile				
Banco Fallabella	Fingerprint	Identix	Live	–
Name of system: BioLogon software and 150 MT Digit optical fingerprint readers from Identix.				
Function: Customer verification during transactions.				
Germany				
Ekey	Various	Various	Discontinued	–
Function: Verifying users over the Internet.				
Ghana				
Bank of Ghana	Fingerprint	TSSI	Live	–
Name of system: VerID fingerprint verification technology from TSSI.				
Function: Staff access.				
Israel				
Bank Hapoalim	Signature	WonderNet	Live	Several million
Name of system: Penflow from WonderNet.				
Function: Customer authentication.				
Luxembourg				
Bank of Luxembourg	Fingerprint	Utimaco and Precise Biometrics	Live	500
Name of system: Single sign on using smart cards and fingerprints from Utimaco and Precise Biometrics.				
Function: Single sign on for banking personnel.				

Table 3.4 *continued*

Name	Technology	Name of supplier	Status	No. enrolled
Netherlands				
GWK	Fingerprint	Precise Biometrics	Live	700
Name of system: Precise 100A fingerprint scanners from Precise Biometrics. **Function:** Secure network logon for employees.				
USA				
McDonalds, Fresno	Fingerprint	Indivos	Pilot	–
Name of system: Pay by Touch from Indivos. **Function:** Secure credit card transactions.				
Progressive Insurance	Speaker	Voicevault	Live	20,000 Next expansion to 100,000
Name of system: Voicevault voice verification. **Function:** Re-set employees' passwords.				
Un-named bank	Fingerprint	Identix	Live	–
Name of system: Identix fingerprint readers (a combination of BioTouch PCMCIA cards and parallel port devices) and a 5000 seat license for Biologon software. **Function:** Network access control				
Various	Fingerprint	Real-Time Kiosks	Live	–
Name of system: Real-Time Kiosks has installed a number of self-service biometric kiosks at Credit Unions throughout the US. These use fingerprint technology from Saflink. **Function:** Customer authentication.				

3.5.3 Government

The governmental market provides some of the most complex challenges for the biometrics industry. At one end, there are companies competing for relatively small verification contracts, which often have similar requirements to those in other vertical sectors (e.g. applications involving access control). At the other end many government departments are now looking seriously at the need to beef up security and reduce the problems of stolen identities through automated authentication processes.

The surge in interest for these large programmes – particularly drivers' licenses, national ID and voter registration – comes at a time when the biometrics industry is undergoing a radical shake-up and consolidation. This is significant for the market as a whole because governmental decision-makers need to be confident that any companies – or consortia of companies – they work with will be able to deliver technology that can adapt and perform well into the long term. Thus, the maturity of the biometrics industry, and the emergence of its industry leaders is an important step for biometrics within this sector.

In spite of positive movements within this sector, however, dramatic 'overnight success' should not be expected. At a senior level in many governments, lip service is currently being paid to addressing the problems of fraud and identification. This can be seen in new legislation, the development of standards and some mission mandate funding. In particular, there is a growing interest in the US regarding the standardisation of drivers' licenses. However, converting interest and earmarked funding into signed contracts is a lengthy process.

It should also be remembered that many large implementations migrate slowly, using a number of different technologies. In such cases, biometrics are often provided as an add-on to existing technologies, such as smart cards. In the case of ID systems, most countries have to switch from paper or plastic-based cards. Therefore any ID rollout is a major step.

3.5.3.1 Summary of Market Applications

The market can loosely be divided into two sections – small-scale programmes and large-scale programmes, which may include identification (1–many searching).

Small-scale programmes cover access control, time and attendance, background checks of employees, network security and tracking of homeless people. Within this section there is also access control to high security environments, such as government-owned nuclear facilities.

Large-scale programmes include military use, voter registration, welfare systems, drivers' licenses and national identification.

3.5.3.2 Drivers

- *Legislation and initiatives* – Legislation and initiatives matched with proper levels of funding provide the strongest driver for growth within the governmental market. Much of this legislation has been spurred on by a global need to improve security in the wake of September 11th. Of particular note, legislation which may benefit the biometrics industry includes the proposed HB1125 which has been suggested to combat identification theft, such as the use of counterfeit licenses to make false credit card and other consumer charges; and the Driver's License Modernisation Act, which aims to standardise drivers licenses. Under the proposed guidelines, the US states and the District of Columbia would have five years to conform to a new nationwide standard for licenses. The US Department of Defense (DOD) Common Access Card project could use a combination of biometrics and smart cards.
- *Standards development* – The development of standards is currently a very strong driver. This makes it easier for systems integrators to handle functionality. It also acts as a good indicator of the maturity of the market and supporting technologies.
- *Lower prices* – Falling prices of biometric devices and software is important to those who are answerable to tax payers. Within the US and Europe, this is a moderate driver in the aftermath of September 11th, as good security is seen as a central objective to many taxpayers. In less developed parts of the

world, however, where national ID and voter registration projects are currently taking place, this is critical as such countries are acutely cost-conscious.

- *Public awareness* – Public interest currently provides a moderate push towards the use of biometrics. People are generally more aware of the limitations of traditional forms of identification, and understand that security often needs to be improved. This makes it easier for governments to push through legislation and implementation of programmes. However, this is only a moderate driver because there is still a lobby of opinion that believes improved security using biometrics could lead to an unjustifiably high loss of freedom.
- *Democracy development* – Democratisation is helping open new doors to the biometrics community. The use of biometrics in voting systems could also form the basis of a national ID system. This could prove important for several African countries that are aware of the need for an ID system, but so far have no robust systems in place.
- *E-government initiatives* – There is a shift in the way people do business with government. Increasingly, citizens are interacting with government departments through electronic means (such as the internet and email). As this number grows, we may see an increased demand for low-cost devices that can authenticate individuals remotely. In the short term, this is not considered to be a major driver. However, in the long term it may prove a profitable avenue for some biometrics vendors.

3.5.3.3 Detractors

- *Scalability* – Many governmental applications require implementation on a massive scale. Such implementations require two things of biometrics companies: 1) the resources to deliver solutions in a timely way; 2) the ability to provide solutions that work regardless of the size of the rollout. This is seen as one of the strongest detractors currently against the use of biometrics. Some potential customers still question the ability of many biometrics companies to be part of a wide-scale rollout. Equally, we often hear of companies who make promises that they cannot or do not keep. This leads to failed projects and a bad name for the industry as a whole.
- *Technical concerns* – Related to the scalability issue, technical concerns are still a fairly big detractor in large-scale projects. In particular, enrolment and integration fears hold some countries back. This is more of a problem in parts of the world where there is a shortage of well-trained people to assist with enrolment. Even before choosing a biometric solution, governments need to consider whether they have enough trained personnel; how they will train additional staff; how sensitive data will be handled; how citizens will be enrolled and so on. Equal to these questions are integration concerns. Countries often want to be able to easily integrate their systems at a local, regional, national and international level.
- *Lengthy decision process* – The slow way in which most governmental departments make decisions is a big problem for many biometrics companies. However, this is also a challenge that is faced by many other industries. This lack of decisive action frequently means that in spite of positive sound-bites from leading politicians, contracts and Requests for Tender, often fail to materialise.

- *Politics* – Politics can be a fairly major detractor to the development of this market. Some people question whether there is actually a will to adequately fund certain projects. This is also matched with the problem that politically expedient solutions are not always the best choice.
- *Cost* – Implementation costs remain a big problem for all government departments. The roll-out of National ID programs – especially in countries like Uganda – can be costly as a percentage of GDP.
- *Privacy* – Privacy concerns are still seen as problems that could hold back the use of biometrics within ID programmes. While there is not too much of a problem in the 1-1 market, privacy advocates, especially in the US and Europe, are deeply concerned about the collection of personal data. In the wake of September 11th, these concerns have subsided somewhat. However, there is no doubt that they will resurface if they are not adequately addressed.

3.5.3.4 Outlook

The biometrics industry is expected to build on its existing foundations to make good progress within all areas of governmental life. September 11th has forced leading governments to think more about security – and also learn about different technologies. The passing of legislation that specifically mentions biometrics is good news to the industry, and – with a little patience – we should see major opportunities for a number of different technologies.

In terms of revenue the market is estimated to reach just over US$500 million by 2006 as revenue from longer-term large-scale contracts begins to appear. Growth throughout the period is expected to be strong, however.

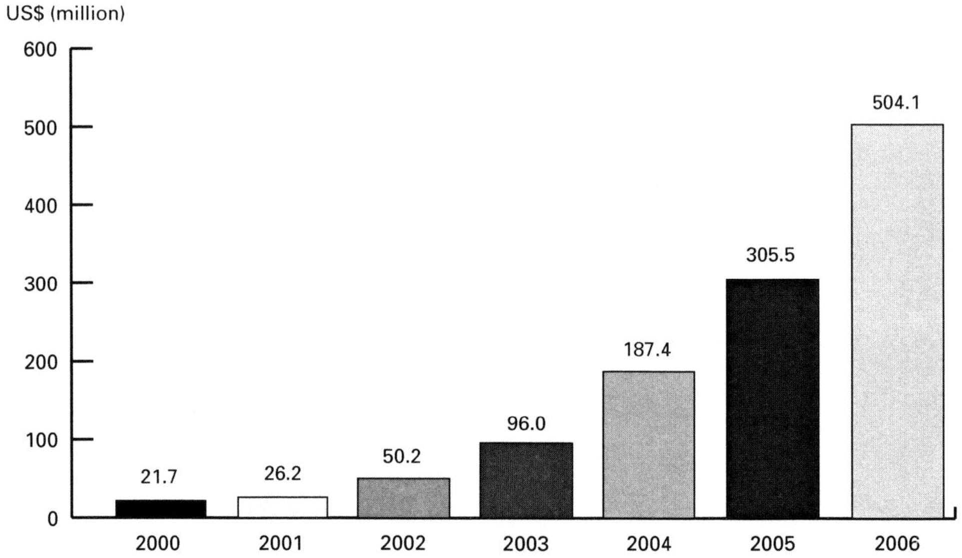

Figure 3.14 Revenue Forecast for the Government Market 2000–2006 (US$ million)

3.5.3.5 A Selection of Government Market Installations

Table 3.5 A Selection of Government Market Installations

Name	Technology	Name of supplier	Status	No. enrolled
Argentina				
ID System	Fingerprint	Printrak	–	–
Name of system: Argentina is believed to have adopted Printrak's biometric technology for its ID system				
Burundi				
ID System	–	–	–	–
The government of Burundi is believed to be installing a national ID system incorporating biometrics.				
Djibouti				
ID System	–	–	–	–
Djibouti is believed to be installing a national ID system incorporating biometrics.				
Republic of El Salvador				
Civil ID system	Fingerprint	Printrak	–	–
Name of system: Fingerprint system from Printrak. Prime contractor of the project is Docusal. **Function:** Identification of 4 million citizens.				
Guatemala				
ID System	Fingerprint	Printrak	–	–
Name of system: Guatemala is believed to have adopted Printrak's biometric technology for its ID system.				
Hong Kong				
National ID scheme	Fingerprint	Cogent	–	–
Name of system: Smart Identity Card System (SMARTICS). The biometric part of the system will use fingerprint technology from Cogent. Cogent is a member of a consortium led by PCCW Business e-Solutions. **Function:** National identification.				
Lebanon				
Lebanese Presidential Palace	Hand	Recognition Systems	Live	–
Name of system: Hand Readers from Recognition Systems. **Function:** Employee access control and time and attendance.				
Lebanese Education, Justice and Information Ministries	Hand	Recognition Systems	Live	–
Name of system: Hand Readers from Recognition Systems. **Function:** Employee access control and time and attendance.				
Macou				
National ID project	–	–	–	–
Name of system: The technology is believed to have been provided by Giesecke & Devrient, Siemens and NEC. **Function:** National ID.				

Table 3.5 *continued*

Name	Technology	Name of supplier	Status	No. enrolled
Mauritania				
Mauritania Voters ID	Fingerprint	Sagem	Live	–
Name of system: Cards incorporating fingerprint technology from Sagem. **Function:** Voter identification.				
Mexico				
Mexican Government	Face	Identix (formerly Visionics)	Live	–
Name of system: FaceIt from Identix (formerly Visionics). **Function:** Voter registration.				
Department of Security, Prevention and Social Readjustment, State of Jalisco	Fingerprint	Identix and Key Tronic	Live	–
Name of system: BioLogon from Identix and fingerprint scanners from Key Tronic. Integrated by Arrasamex. **Function:** Network security.				
Spain				
TASS Project	Fingerprint	Unisys and Identix (formerly Identicator)	Live	Over 500,000
Name of system: Optical fingerprint readers combined with smart cards. **Function:** Complete social security application allowing users to verify themselves at unattended kiosks.				
Uganda				
Uganda Voters ID	Face	Viisage	Live	8 million
Name of system: Facial recognition system from Viisage Technology and character recognition system from Readsoft UK. **Function:** Voter identification.				
USA				
California	Fingerprint	–	–	–
Function: Uses fingerprints for driver license registration.				
Georgia	Fingerprint	–	–	–
Function: Uses fingerprints for driver license registration.				
Hawaii	Fingerprint	–	–	–
Function: Uses fingerprints for driver license registration.				
Oklahoma	Fingerprint	–	–	–
Function: Uses fingerprints for driver license registration.				
Texas	Fingerprint	–	–	–
Function: Uses fingerprints for driver license registration.				

Table 3.5 *continued*

Name	Technology	Name of supplier	Status	No. enrolled
USA *continued*				
Unknown county	Fingerprint	Digital Descriptor Systems	–	–
Name of system: Fingerprint Matching System (FMS) from Digital Descriptor Systems. **Function:** Identifying and tracking of homeless people by the county's Department of Health and Human Services (HHS).				
US Army	Face	Viisage	–	–
Name of system: Face recognition technology licensed from Viisage. **Function:** Access control and other high security deployments.				
West Virginia Dept of Motor Vehicles	Face	Digimarc	Pilot	–
Name of system: FaceIt technology from Identix (formerly Visionics), implemented by Digimarc ID Systems. **Function:** Drivers license registration.				

3.5.4 Healthcare

As pressures increase on health care providers to cut fraud, improve physical and digital security and reduce costs, biometrics are beginning to be seen as a potential force within the healthcare sector. Driven on by legislation in the US, and increasingly automated procedures in other parts of the world, this is an increasingly important avenue for a wide variety of biometric technologies.

The sheer breadth of requirements from healthcare providers illustrates clearly that 'one size does not fit all'. The implementation of biometrics for access control, network access control and general healthcare duties demonstrates a need for a number of different types of technology, and a range of biometrics are being used to address these needs.

3.5.4.1 Summary of Market Applications

The following is a list of typical biometric applications in the healthcare sector:

- Network access control
- Centralised and decentralised systems
- Drug control
- Healthcare for specific diseases (such as Parkinson's) – often smart card using fingerprint and containing prescription info.
- Access to medical records
- Access control to buildings

3.5.4.2 Drivers

- *HIPAA legislation* – In the US, the rush to meet HIPAA compliance deadlines is typically seen as a major driver for biometrics. We believe that this is a significant piece of legislation, which has been passed at a time when biometric technology is reaching maturity. For this reason, there is a growing interest in the use of a wide variety of biometric technologies – often in conjunction with other methods of authentication (e.g. smart cards). However, biometrics do not address all of the challenges of HIPAA, so biometrics companies are still competing against other – more established – technologies.

- *Increased reliance on IT systems* – The growing use of computers and Internet-based technology throughout the healthcare sector, for applications such as accessing patient records, provides a potentially large driver for the biometrics industry.
- *Privacy enhancement* – Healthcare providers increasingly recognise that they must address security and privacy concerns. This often means that they are looking for technologies that provide audit trails of people accessing records; protect patient records; reduce fraud and simplify administrative procedures. Such systems could offer two benefits: reduce the amount of money lost through fraudulent use of healthcare systems; and ensure that the right patient receives the right treatment. This is considered a moderate strength driver as cost savings may not be huge.
- *Remote access* – As more people travel and move overseas, there is a growing need to access individual patient records remotely. Although this is not a significant driver today, there remains an interest in finding a solution that will address this need in the future.

3.5.4.3 Detractors

- *Competing technology* – The biggest detractor to the use of biometrics is competition from other technologies, such as smart cards. This is particularly true in the case of HIPAA compliance, where biometrics are competing against other – more established – technologies.
- *Slow decision-making* – The second biggest detractor is the lengthy decision-making process and reluctance by many healthcare professionals to 'take ownership' of a project. This – coupled with a lack of budget – means that the technology is progressing at a much faster rate than the decision.
- *Misperceptions* – Misperceptions and expectations of biometrics and their applications is still a strong detractor. These misperceptions include end-user concerns about what biometrics can do and how they work. Confusion about biometrics can be seen amongst leaders within the healthcare industry and also within the consulting community. These deficiencies lead to a reliance on older, more understood, technologies like passwords.
- *Cost constraints* – Most organisations within the healthcare sector operate on very tight margins: this means that cost is still a moderately strong detractor. Unlike time and attendance, increased security is not necessarily a cost-effective process, although opinions are changing since September 11th.
- *IT issues* – Although the healthcare sector's use of IT is growing, this is still an industry which is not particularly IT-savvy. This acts as a moderate detractor as it limits the number and types of application that biometrics can be applied to.

3.5.4.4 Outlook

Many healthcare providers are positive about the potential for biometrics within the healthcare market. They know that identification and verification problems need to be addressed and stamped out. But if things are to take off, a lot of organisations need decisions to be taken at a high – often governmental – level. Governments need to show themselves to be interested and cooperating with the technology. In the case of state-run health services, these governments also need to demonstrate a real financial commitment to addressing the problems of fraud, privacy, security and patient records.

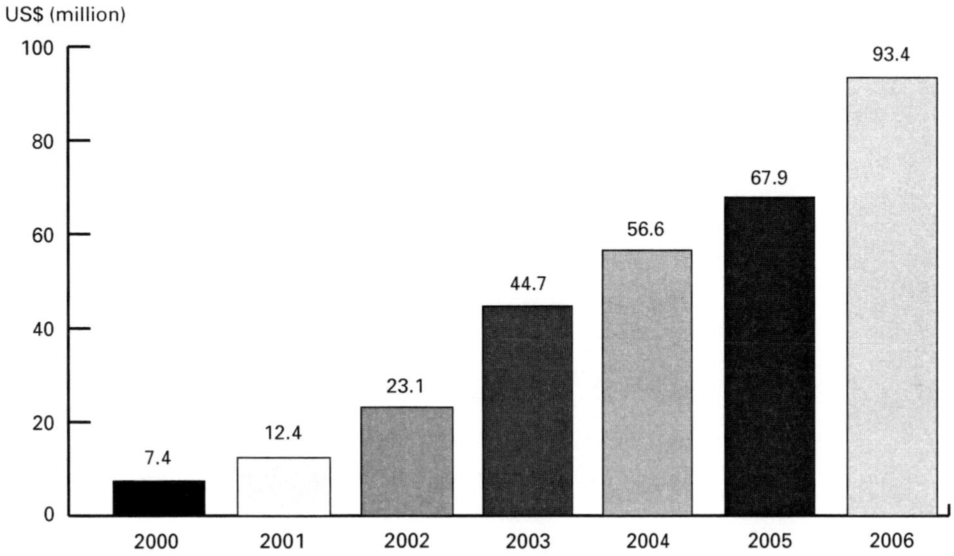

Figure 3.15 Revenue Forecast for the Healthcare Market 2000–2006 (US$ million)

Without doubt HIPAA regulations will drive this market over the next few years, but in the subsequent years this market should still provide excellent growth potential, as long as prices fall and knowledge of the technology improves. Revenue between 2002 and 2003 will grow from US$23 million to almost US$45 million. By the end of the forecasting period in 2006 growth will have accelerated to generate US$93 million.

3.5.4.5 A Selection of Healthcare Market Installations

Table 3.6 A Selection of Healthcare Market Installations

Name	Technology	Name of supplier	Status	No. enrolled
Netherlands				
Dutch Burns Foundation	Speaker and Fingerprint	Keyware	–	–
Name of system: Uses layered biometric technology from Keyware, PKI from Baltimore Technologies, Digital Certificates from DigiNotar and smart cards from Gemplus.				
Function: Access to burns patient records.				
IVZ, Ministry of Justice, Ministry of Healthcare	Biometrics and smart cards	HSB Cards & Card Systems	–	–
Name of system: Uses smart cards from Infineon and Card Care card management system, application software, application hardware from HSB Cards.				
Function: Identification of addicts.				
Zorg en Zekerheid Health Insurance Company	Fingerprints	HSB Cards & Card Systems	–	–
Name of system: Smart cards with fingerprint sensors from HSB Cards.				
Function: Identification and prescription information of Parkinson's Disease sufferers.				
UK				
Unknown hospitals	Fingerprint	ISL Informer Systems	Pilot	–
Name of system: Informer Systems claims to have 55 pilots throughout the UK using its SentriNET base engine including some with e-SentriNET.				
Function: Network security.				
USA				
Antelope Valley Hospital/IT HIPAA Readiness project	Fingerprint	Siemens and BioconX	Live	2200
Name of system: Network security software from BioconX and fingerprint sensors from Siemens.				
Function: Biometric single sign on and provide audit trail of users.				
Catholic Healthcare West	Fingerprint	Identix	Pilot	–
Name of system: Fingerprint readers and Biologon software from Identix.				
Function: Secure remote physician access to patient records.				

Table 3.6 *continued*

Name	Technology	Name of supplier	Status	No. enrolled
USA *continued*				
Eagleville Hospital	Iris	Politec, Iridian and Panasonic	Live	100. Rollout is planned for 250.
Name of system: Politec Authentication Security Suite (PASS) software from Politec, which uses iris recognition technology combined with the Panasonic Authenticam desktop camera. **Function:** Single sign on. The system is being used to control log on to local computers and then access confidential information that is stored on Eagleville's Wide Area Network. PASS is also being integrated into Eagleville's primary healthcare system – QuadraMed – to secure access to patient medical information.				
Inova Fairfax Hospital	Fingerprint	Bioscrypt, eko systems	–	–
Name of system: Adopted system from Bioscrypt and eko systems. **Function:** Control access to peri-operative medical records.				
New York State Office of Mental Health	Fingerprint	Biometric Access Corp	–	–
Name of system: Secure Touch from Biometric Access Corp. **Function:** Protect patient records.				
St Vincent Hospitals and Health Care Centers	Fingerprint	Computer Associates and Saflink	Live	–
Name of system: Fingerprint sensors supplied by AuthenTec. Systems supplied by Computer Associates and Saflink. **Function:** Network security.				
Scott & White Memorial Hospital	Fingerprint	Identix	Live	–
Name of system: Identix BioTouch PC Fingerprint Cards with BioLogon authentication software. **Function:** Network access control for employees.				
WEN Extended Care Facilities Management Corp	Hand	Recognition Systems	Live	–
Name of system: Hand Punch from Recognition Systems. **Function:** Employee time and attendance.				

3.5.5 Law and Order

The biometric industry has built strong foundations in the law and order market – and these are set to continue in the post-September 11th world. In the area of law enforcement, fingerprint, AFIS and face technologies have been involved in numerous pilots and contracts. While the results of a few of these pilots have not always been completely satisfactory, there is a general feeling of optimism that this is an industry where biometrics can make a real impact. Biometrics are also revealing themselves to be good technologies in correctional environments where there is a need to move high volumes of prisoners and visitors securely. In particular, we are seeing hand recognition technologies performing well in this environment.

3.5.5.1 Summary of Market Applications

The following is a list of typical biometric applications in the law and order sector:

- Criminal record clearance/background checks (where livescan devices or AFIS systems are used, these are not included in the revenue predictions for this report)
- Surveillance
- Access control to prisons, court houses etc
- One to many searches of large criminal databases
- One to one searches to identify people such as prison inmates and their visitors
- Offender tracking
- Identification of individuals through mobile devices, for example at the roadside or at a port of entry
- Booking systems
- Network security
- Identification of victims of crime – e.g. victims of paedophiles
- Smart guns

3.5.5.2 Drivers

- *September 11th* – The biggest driver to the use of biometrics is the impact of September 11th on both decision-makers and the public in general. The need to stamp out identity theft and the increased desire for security has provided a surge in interest in all technologies – particularly fingerprint and face.
- *CCTV market growth* – The rising number of CCTV cameras already installed in many parts of the world, provide the basic infrastructure from which to develop biometrically-enabled surveillance systems. This has the potential for strong growth, provided that privacy concerns can be addressed.
- *Freeing up resources* – The need to free up police resources provides a major driver within the law and order market. The largest, and most mature, part of this market continues to be AFIS technology and increasingly livescan systems, which can be used remotely to capture forensic-quality fingerprints for subsequent use in an AFIS system (both AFIS and livescan devices are not included in this report's revenue assessment statistics, however). In addition to burgeoning livescan orders, increased interest in automated mug shot processing will provide significant revenue to the industry.

- *Prison overcrowding* – Overcrowding in prisons has accelerated the need for good offender tracking systems. This is a moderately strong driver as other forms of authentication – such as tagging – are currently receiving a high amount of attention.
- *Secure access control requirements* – Many parts of the law and order market – particularly prisons – need an access control system where high volumes of people can be moved securely. This will continue to provide a moderate driver to suppliers of robust technologies such as hand recognition well into the future.

3.5.5.3 Detractors

- *Privacy concerns* – The biggest detractor within certain parts of the law and order market is the question of privacy. Although privacy questions are more diluted than they were before September 11th, there is no doubt that this is challenge that the industry must address if it is to progress further, particularly in the area of surveillance.
- *Technical and reliability issues* – Technical concerns and reliability are another big detractor. These problems can be seen in the relative immaturity of some biometric technologies, poor integration (e.g. poor lighting, bad placement of cameras etc) and inadequate training. In addition to this, questions about template storage need proper consideration: Should an individual carry their template or should it be held in a database? If held in a database, organisations need to examine questions of privacy and database sizes. In a large city with a large database of templates, performance of the system is a major concern.
- *Unfeasible expectations* – Unrealistic expectations are a major detractor. In the months following September 11th, there was frenzied interest in biometrics as the answer to many security problems. However, law and order agencies need a clear understanding of the performance capabilities of biometrics. People need to know that biometrics have weaknesses – just as other methods of authentication have weaknesses. Companies that make promises and claim otherwise are giving the whole industry a bad name, and could be providing long-term damage to perceptions and growth of biometrics.

3.5.5.4 Outlook

Numerous opportunities exist within this market. While there will continue to be a solid number of AFIS and mugshot management contracts won, considerable interest in other areas of business, such as surveillance, criminal records clearance and offender tracking will be seen in 2003. This, backed up by further contracts for access control devices within correctional institutions, will help the biometric industry to rapidly increase its revenue within the law and order domain reaching US$280 million by 2006 from US$33 million in 2002.

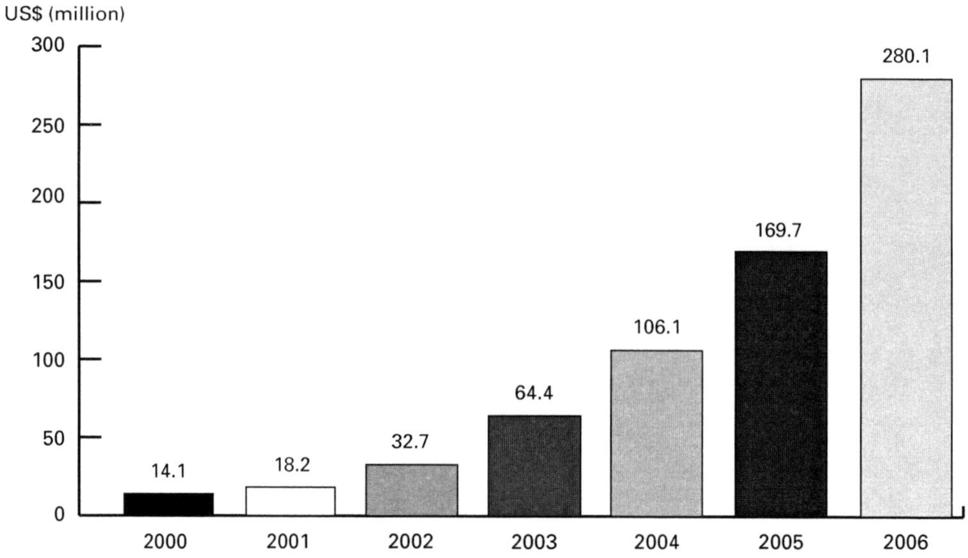

Figure 3.16 Revenue Forecast for the Law and Order Market 2000–2006 (US$ million)

3.5.5.5 A Selection of Law and Order Market Installations

Table 3.7 A Selection of Law and Order Market Installations

Name	Technology	Name of supplier	Status	No. enrolled
Brazil				
Brazilian High Court of Justice	Fingerprint	Identix	Live	–
Name of system: Biologon from Identix.				
Function: Network access control.				
Philippines				
NBI Clearance Processing Project	Face and Fingerprint	Identix (formerly Visionics) and Identix	Live	2 million
Function of system: Criminal record clearance from 15,000 applicants per day. The system uses facial recognition as a primary search engine and fingerprint identification for 1:1 identification.				
Republic of Slovenia				
National Police	AFIS/Palmprint	Printrak	–	–
Name of system: Omnitrak 8.0.1 system from Printrak.				
Function: Capture, store and search over 265,000 fingerprint and palmprint images				

Table 3.7 *continued*

Name	Technology	Name of supplier	Status	No. enrolled
South Africa				
–	Speaker	Voicevault (formerly Buytel)	–	–
Name of system: Group 4 Falck will use Voicevault's Teletrack/Corrections service.				
Function: Tracking of offenders who have been released into the community.				
Switzerland				
Swiss Federal Office of Police	Fingerprint	Printrak	–	–
Name of system: Omnitrak 8.0 AFIS/Palmprint Identification Technology from Printrak.				
Function: Identification.				
UK				
Home Office – various sites	Face, Speaker	Identix (formerly Visionics) and OnGuard Plus (OGPL)	–	–
Name of system: Face recognition from Identix (formerly Visionics). OGPL is working with Keyware to provide a system called Vqvoice to Securicor Custodial Services, a supplier of electronic monitoring services to the programme.				
Function: Tracking persistent young offenders.				
London Borough of Newham	Face	Identix (formerly Visionics)	Live	–
Name of system: FaceIt technology from Identix (formerly Visionics).				
Function: Surveillance.				
NAFIS Project	Fingerprints	TRW and PITO	Live	–
Name of system: NAFIS developed by Police Information Technology Organisation (PITO) and contracted by TRW.				
Function: Identification of crime scene marks.				
National Crime Squad	Face	Imagis Technologies	–	–
Name of system: ID2000 facial recognition from Imagis.				
Function: Identification of victims of child abuse.				
Operation Safer Streets (West Midlands Police)	Face	Identix (formerly Visionics)	Pilot	–
Name of system: FaceIt from Identix (formerly Visionics).				
Function: Surveillance.				

Table 3.7 *continued*

Name	Technology	Name of supplier	Status	No. enrolled
USA				
Alameda County, California	Face	Imagis Technologies	–	–
Name of system: ID-2000 facial recognition combined with Computerized Arrest and Booking System (CABS) from Imagis.				
Function: Digital image and data sharing system for use by all law enforcement agencies throughout the county.				
Barnstable County Sheriff's Dept	Iris	Tier Technologies	Live	1200
Name of system: Tier uses Iridian's iris recognition technology to create a solution that is integrated with IMATS jail management system.				
Function: Identification of inmates, employees and visitors.				
Dept of Justice	Fingerprint	Identix	–	–
Name of system: Joint Automated Booking System contract won by PEC Solutions. The system uses Identix and Cross Match Live Scan devices for capture, and compression software from Aware				
Lake County, Florida	Iris	Tier Technologies	Live	1000
Name of system: Tier uses Iridian's iris recognition technology to create a solution that is integrated with IMATS jail management system.				
Function: Identification of inmates, employees and visitors.				
Oakland Bay Area Rapid Transit	Fingerprint	Identix	–	–
Name of system: Live Scan TouchPrint 2000 Applicant Fingerprint System.				
Roanoke County Sheriff's Dept	Iris	Tier Technologies	Live	1500
Name of system: Tier uses Iridian's iris recognition technology to create a solution that is integrated with IMATS jail management system.				
Function: Tracking movements within facilities and identifying inmates, employees and visitors.				
SQUID / Ontario Police Department	Fingerprint	Identix	Pilot	–
Name of system: IBIS mobile identification system from Identix.				
Function: Remote identification of suspects.				
Tampa	Face	Identix (formerly Visionics)	–	–
Name of system: FaceIt from Identix (formerly Visionics).				
Function: Surveillance.				

Table 3.7 *continued*

Name	Technology	Name of supplier	Status	No. enrolled
USA *continued*				
Tennessee Bureau of Investigations	Fingerprint	Identix	Live	–
Name of system: Fingerprint scanning by Sylvan/Identix Fingerprinting Centers.				
Function: Employee background checks.				
Virginia Beach City Council	Face	Identix (formerly Visionics)	Live	–
Name of system: FaceIt from Identix (formerly Visionics).				
Function: Surveillance.				

3.5.6 Leisure

Like the financial services market, many parts of the leisure industry revolve around customer satisfaction. For this reason, any technology used in customer-facing areas must provide convenience over security.

In the casino environment, biometrics are being successfully utilised in a number of applications, especially surveillance and cash access. Generally, decision-makers within casinos are keen to embrace new technologies – and biometrics have benefited from this. Many parts of the leisure industry also require fast processing of a large number of customers (e.g. football clubs and theme parks). Pilots have taken place within these areas, but have so far had mixed results. This part of the market could develop further with the use of more robust technologies.

3.5.6.1 Summary of Market Applications

The following is a list of typical biometric applications in the leisure sector:

- Access control to facilities
- Surveillance of individuals within casino environments
- Cash access in casinos
- Sports matches
- Ticketing – e.g. football fan tickets
- Frequent user passes to theme parks
- Night club entry

3.5.6.2 Drivers

- *Positive attitude* – One of the biggest drivers within this market is the positive attitude that many companies have towards the use of biometrics. In customer-facing areas, the technology must be both fun and convenient. In back-office functions, the emphasis is on security. Biometrics have made a good start in both these areas, and we can expect to see them used in additional areas such as in-room gaming and player identification. We also anticipate further interest from theme parks and health clubs for applications such as frequent user access.

- *Casino market* – This is a fast-moving market for biometrics, in particular to identify undesirable patrons using facial recognition. However, another strong driver is the ongoing need for cash by users of casinos. One answer is to provide biometric platforms in casinos' ATMs and cages. This provides casinos with a footprint that will enable them to underwrite and reduce the risk of transactions: There is usually a daily limit to the amount of money that any individual can take from an ATM – by using biometrics, casino ATMs can allow patrons to exceed that limit. The use of automated cash withdrawal procedures is important as it appeals to the casinos' need to take advantage of impulse actions. The psychology of taking money from a machine is considered more favourable than frequently visiting a cashier to get more money.

3.5.6.3 Detractors

- *Privacy concerns* – The biggest detractor remains the perception of privacy invasions. While this is less of an issue in back office areas, it is essential that end-users are comfortable with how their personal data is collected and stored when used in customer-facing operations.
- *Technological capability in difficult environments* – Technology questions provide another detractor. Casinos usually operate in noisy conditions with variable lighting. These conditions are not always ideal for biometrics such as face recognition and speaker verification – so it is essential that integrators take adequate account of such factors during implementation. Many parts of the leisure industry are concerned with processing large volumes of people as quickly as possible. Therefore, it is also essential that companies choose a technology that is robust enough to handle large groups of people.

3.5.6.4 Outlook

The leisure industry provides many opportunities for future development. Behind the scenes activities, such as time and attendance and access to vaults, will continue to provide a good-sized market within this sector. It can be expected to see more instances of biometrics used with patrons of places such as casinos – provided that the technology is seen as convenient and acceptable.

In terms of revenue there was very strong growth between 2001 and 2002 as the casino sector produced a spate of deals. By 2006 the sector will only still form a small share of the overall market at just under US$19 million, but will nevertheless be growing strongly.

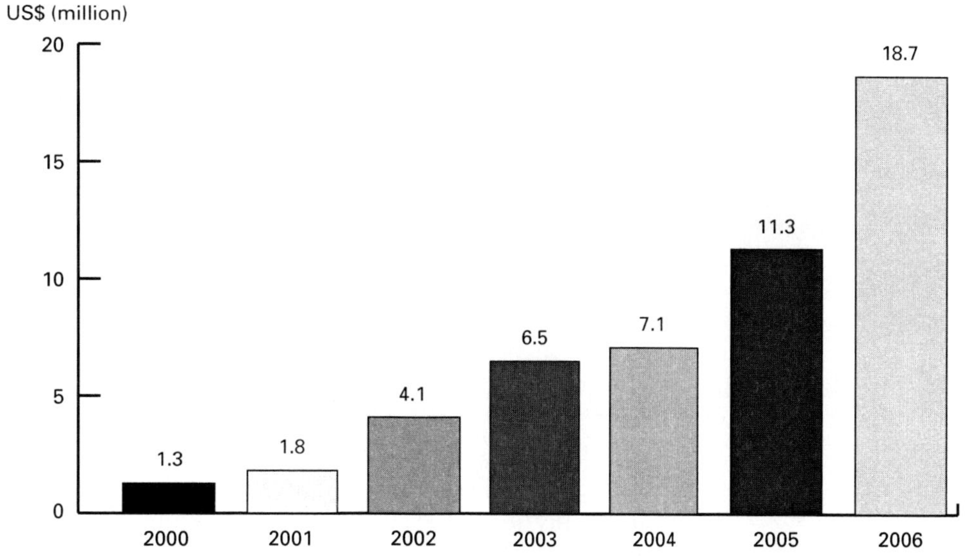

Figure 3.17 Revenue Forecast for the Leisure Market 2000–2006 (US$ million)

3.5.6.5 A Selection of Leisure Market Installations

Table 3.8 A Selection of Leisure Market Installations

Name	Technology	Name of supplier	Status	No. enrolled
USA				
Arrow Head Pond	Hand	Recognition Systems	Live	–
Name of system: HandKey II from Recognition Systems. **Function:** Access control.				
US Super Bowl, Tampa	Face	Viisage	–	–
Name of system: Face recognition from Viisage. **Function:** Surveillance.				
Various gaming outlets	Face	Viisage, Global Cash Access and Infonox	Pilot	–
Name of system: Automated Cashier Machine using technology from Viisage, Global Cash Access and Infonox. **Function:** Cash withdrawal within casinos.				
Various US casinos	Face	Viisage	–	–
Name of system: Face recognition from Viisage. **Function:** Surveillance Information Network, which connects all member casinos, so that information, including facial images, can be exchanged between casinos and retrieved for future use.				

3.5.7 Telecoms

Biometrics have so far failed to make much of an impact in the telecoms market. While manufacturers are still praising the benefits of biometrics within this market, most work remains at the developmental stage. The main reason for the slow growth would appear to be the lack of interest in security by many telecommunications companies. This comes in spite of pushes from governments and consumer groups to reduce the level of street crime associated with the ease of utilising a stolen mobile phone.

The main telecoms market is still in the area of business-to-customer transactions, and there is a general belief that the industry is at least two years away from generating enough revenue or large-scale systems to justify a major spend on infrastructural security. Many telecoms companies are keeping their eyes open to the possibility of biometrics: in particular cost-per-user ratios, third party authentication, privacy and data legislation. However, at this moment in time, apart from some installations, the industry is doing little more than paying lip service to the potential for the technology.

A few small pockets of interest do exist within this market: notably in the area of satellite phones. Companies operating in this sector seem interested in reducing the amount of fraud that many of them are currently suffering with existing PIN and password methods. One answer to this question could be the use of biometric technology.

3.5.7.1 Summary of Market Applications

The following is a list of typical biometric applications in the telecoms sector:

- Mobile phone security
- Accessing phone company services, such as phone credit top up systems
- Satellite phones
- Securing 3G services
- Physical security of telecoms infrastructure

3.5.7.2 Drivers

- *Device security* – The biggest driver is currently the need to improve and provide security of devices such as mobile phones. Biometrics could be used to counter fraud and offer repudiation that a person using a telephone is who he or she claims to be.
- *3G services security* – In the medium to long term, further interest will be seen in securing new services. Such services will likely be provided by 3G systems, so little take up for at least two to three years is expected.
- *IT/telecoms convergence* – The convergence of IT and telecoms is having an impact on every area of business and domestic life. This provides additional opportunities for biometrics, although there is a tremendous amount of competition from other forms of authentication.

3.5.7.3 Detractors

- *Lack of business case* – The biggest detractor is the lack of a clear business case that shows how biometrics will reduce costs for the telecoms provider. This is essential for two reasons: (1) most telecoms providers have little interest in security (2) many companies have invested heavily in 3G technology and are reluctant to divert funds to other aspects of business.
- *Competition* – The second biggest detractor is very strong competition from other forms of authentication – notably PIN. This will remain the case for a considerable amount of time.
- *3G delay* – Many decisions are not expected to be made until the rollout of 3G is complete. This is stalling investment and many analysts believe that there will be a delay of at least two years.
- *Customer reaction* – As the technology is new and largely untried, many manufacturers are unwilling to try biometrics until they are certain of their 'mass appeal'.

3.5.7.4 Outlook

Biometrics certainly have the potential to develop well within the telecoms market place. In particular, fingerprint and speaker verification technology could prove useful forms of authentication. However, until succinct business cases – which show a definite advantage to telecoms operators – can be made, there will be moderate short-term growth within this sector. The greatest opportunities for biometrics will come in the long term – when telephone companies start taking a serious look at how to reduce fraud. By 2006 revenue is expected to be US$28 million from US$8.9 million in 2002.

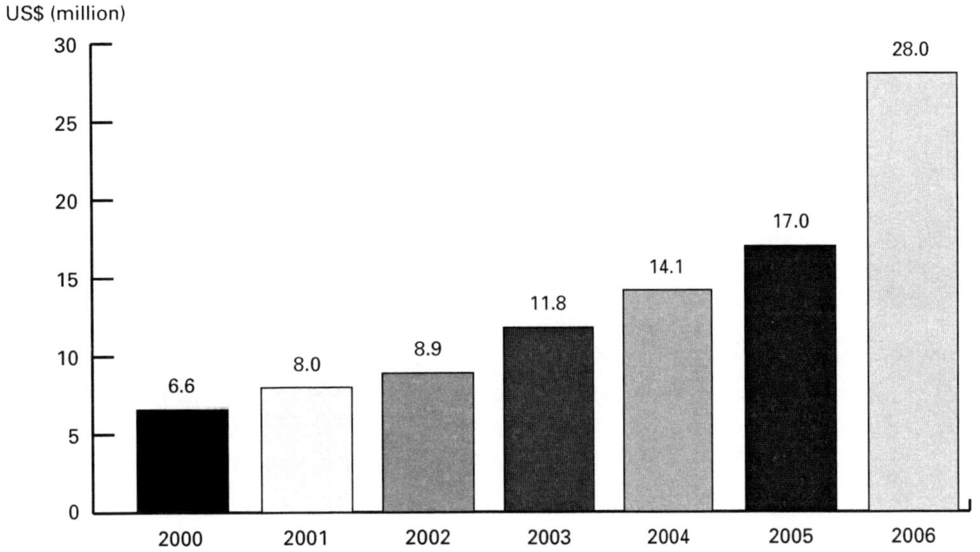

Figure 3.18 Revenue Forecast for the Telecoms Market 2000–2006 (US$ million)

3.5.7.5 A Selection of Telecoms Market Installations

Table 3.9 A Selection of Telecoms Market Installations

Name	Technology	Name of supplier	Status	No. enrolled
Major global telecoms company	Speaker	Voicevault (formerly BuyTel).	Live internally	12000
Name of system: Voicevault Voice Verification. **Function:** Automated Voice Portal Access to Information Services, Secure Access Applications and Payment Applications (Bill Payments etc).				
T5 PrePay	Speaker	Trust5 and Nuance	Pilot	250
Name of system: T5 Risk Algorithm and Nuance Verifier. **Function:** Pre-paid top-up application.				
Trium Mondo GSM/ GPRS PDA phone (Mitsubishi Electric Telecom).	Speaker	Schlumberger-Sema and Domain Dynamics	Pilot	–
Name of system: The system is provided as a SIM Tool Kit applet (from SchlumbergerSema), which can be loaded onto any standard Java Card SIM. **Function:** Safeguard access to mobile phones.				

3.5.8 Travel and Immigration

Even before September 11th, the travel and immigration market was proving a popular outlet for biometric technologies. Since the terrorist attacks, this has become an increasingly important sector – although a change in emphasis has occurred. Previous to the third quarter of 2001, the question from airports was one of passenger processing speed. Now the challenge has shifted to security. The industry has been the subject of media hype, which has been good for some companies in the short-term. However, long-term development still requires education about how the technology works, so as to avoid disappointing failures.

The use of biometrics within immigration applications is one area that is enjoying a considerable amount of success in Europe – particularly with asylum seekers. A European asylum seeker identification programme, Eurodac, is due to go live in January 2003. There is also an on-going need to identify people and provide cards that can be used by immigrants. Significant laws have been passed in Italy about all immigrants (not just asylum seekers) carrying a form of ID. Meanwhile, Switzerland has also updated its systems.

Many applications involving face and fingerprint technologies for identification purposes are expected. However, in instances where large quantities of people need to be processed quickly, further contracts for iris technologies are likely. There could also be a number of schemes that use a combination of biometrics, such as hand and facial recognition, so as to avoid problems associated with failure to enrol rates.

3.5.8.1 Summary of Market Applications

The following is a list of typical biometric applications in the travel and immigration sector:

- Frequent traveller passes
- Immigration controls
- Processing of asylum seekers
- Access control/time and attendance of airport personnel
- Smart card ticketing systems incorporating biometrics
- Potential access control to planes and cockpits
- Efforts to stop tailgating
- Biometric-enabled travel documents
- Surveillance
- Automated border passenger systems
- Airport security
- Employee background checks
- Network access control

3.5.8.2 Drivers

- *September 11th* – The aftermath of September 11th is providing the strongest driver for biometrics within this market. This can be seen in an increased desire for surveillance and other security devices.
- *New legislation* – Legislation, such as the *'Enhanced Border Security and Visa Entry Reform Act of 2001'* and the *'Legal Immigration and Family Equity Act'* show a governmental commitment to authentication. This is a very strong driver, particularly when biometric technology is specifically mentioned within the act.
- *Increased travel across borders* – The growing number of people crossing borders or making frequent border crossings is a major driver. Such people need identifying and processing as quickly and securely as possible. Many frequent travellers are also prepared to pay for a system that will speed up their progress through border control.
- *Asylum seekers* – The increased number of asylum seekers in Europe provides certain processing challenges to many governments. This is helping to drive the use of biometrics through programmes such as Eurodac.
- *Implementation history* – The success of past projects, such as the Ben Gurion Airport scheme in Israel, provide evidence that biometrics can work within the travel environment, as long as the system is carefully implemented. This is a fairly important driver as it reinforces the fact that biometric technology is reaching maturity.

3.5.8.3 Detractors

- *Education* – The biggest detractor within the travel and immigration market will probably be education. Since September 11th, many agencies have had a knee jerk reaction to overcoming the problems of further terrorist threats. This sometimes means that they have sought the wrong technology for the job – which could ultimately lead to failure. Pilot projects are also sometimes undertaken without setting out a clear objective. Without a stated objective, bad results – and bad publicity – may occur. As part of the education process, it is important that potential customers understand the strengths and weaknesses of biometrics versus the strengths and weaknesses of other technologies. This is particularly true in the case of surveillance, where critics have been concerned that devices offering less than 100% security may take up too much time to address all the false matches that such a system could generate. This doesn't take into account the fact that no device is 100% secure. It also assumes that a system's primary function is to lead to instant apprehension.

- *Privacy concerns* – A powerful privacy lobby that fears personal harassment and the wrongful identification of suspects is also a major detractor. This was a very serious concern before September 11th, and will continue to resurface throughout the short term. Some of these groups question a perceived invasion of privacy in relation to the effectiveness of some technologies. While people may be forced to put up with such issues in the immigration market, they won't put up with them in travel programmes.
- *Scalability* – Scalability questions are a fairly major detractor. The logistics of creating large databases of people expected to travel through a port are enormous. Added to this, some authorities are concerned about how easy it is for the general population to use certain technologies. There are also concerns about whether devices are robust enough to cope with high volumes of passenger numbers.
- *Standards* – In many large-scale rollouts, compatibility with existing databases, interoperability and travel document standards still need to be addressed.
- *Business case* – In the case of frequent traveller programmes, the business case for biometrics may be questionable as frequent travellers make up a very small percentage of the travelling public.

3.5.8.4 Outlook

The travel and immigration market will provide a good source of revenue for the biometric industry throughout the short, medium and long term. Revenue growth in this market will be fast, moving from US$16 million in 2002 to US$224 million by 2006.

The global need to provide secure identification, combined with a growing demand to automate passenger processing, will aid the development of biometrics within this sector. The increased use of biometrics in machine readable travel documents will occur, and the increased interest in airport security in general will continue. Immigration controls and asylum seeker programmes will also continue to grow. Airline cockpit doors and controls could be covered by a biometric device – although this is not an area that will grow for some time.

If the industry is to maximise the opportunities within this market into the long-term, it is essential that privacy and educational concerns are fully considered.

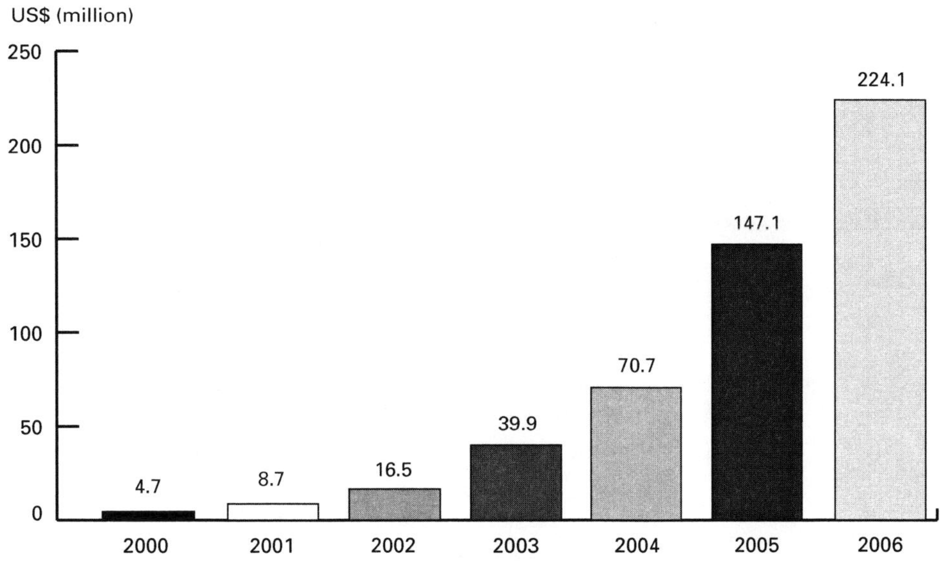

Figure 3.19 Revenue Forecast for the Travel and Immigration Market 2000–2006 (US$ million)

3.5.8.5 A Selection of Travel and Immigration Market Installations

Table 3.10 A Selection of Travel and Immigration Market Installations

Name	Technology	Name of supplier	Status	No. enrolled
Australia				
Sydney Airport	Face	–	Pilot	–
Function: One-to-one check of passports against faces.				
Sydney Airport	Face	–	Pilot	–
Function: One-to-many pilot that uses surveillance cameras to identify wanted faces within the airport environment.				
Canada				
–	Hand	Recognition Systems	Live	–
Name of system: OCR card combined with ID3D card from Recognition Systems. **Function:** Border control.				
Guatemala				
–	Civil Fingerprint ID	Printrak	–	–
Name of system: Printrak Civil AFIS. **Function:** Passport authentication and identification.				

Table 3.10 *continued*

Name	Technology	Name of supplier	Status	No. enrolled
Iceland				
Keflavik Airport	Face	Identix (formerly Visionics)	Live	–
Name of system: FaceIt technology from Identix (formerly Visionics).				
Function: Border control and surveillance.				
Israel				
Ben Gurion Airport Authority	Hand	Recognition Systems	Live	–
Name of system: Express Entry from EDS. This is combined either with an Israeli bank credit card or with a pre-printed card provided by Ben Gurion Airport.				
Function: Automated passport control.				
Italy				
A Milan airport	Face	–	Pilot	–
Function: Frequent flyer initiative that – subject to confirmation – will involve the use of face recognition and smart cards to be used for secure processing of frequent travellers. The initiative was announced by IATA and SITA.				
Malaysia				
Kuala Lumpur International Airport	Finger	–	–	–
Function: Border control.				
Netherlands				
Alien Police, Rotterdam	Iris	Joh. Enschede	Pilot	250
Name of system: Iris recognition technology provided by Joh. Enschede.				
Function: Reduce fraud and paperwork associated with immigration.				
Schipol Airport	Iris	–	–	–
Function: Automated passenger boarding and quick security access.				
Peru				
Instituto de Immigracion	Hand	Recognition Systems	Live	–
Name of system: ID3D from Recognition Systems.				
Function: Border crossing and passport control.				
Switzerland				
Zurich Airport	Face	–	Pilot	–
Function: Frequent flyer initiative that – subject to confirmation – will involve the use of face recognition and smart cards to be used for secure processing of frequent travellers. The initiative was announced by IATA and SITA.				

Table 3.10 *continued*

Name	Technology	Name of supplier	Status	No. enrolled
UK				
Gatwick Airport	2 finger geometry	Bio-Met Partners	Pilot	–
Name of system: Combined biometric/smart card trial using Digi-2 from Bio-Met Partners and integrated by Fortress GB. **Function:** Processing of frequent travellers.				
Heathrow Airport	Iris	EyeTicket	Pilot	–
Name of system: Jetstream Passenger Processing from EyeTicket. **Function:** Passenger processing of frequent travellers as part of the IATA SPTIG.				
UK Immigration and Nationality Directorate	–	–	–	–
Function: 6-month voluntary repatriation project for Afghan asylum seekers. It is believed that those eligible will be provided with money and a flight back to Afghanistan. Biometrics will be used to verify the identity of recipients.				
UK Immigration and Nationality Directorate	AFIS	Sagem	Live	–
Name of system: Livescan system from Sagem. **Function:** Fast identification of asylum seekers.				
UK Immigration Service	AFIS	Sagem	Live	–
Name of system: Cardscan system from Sagem. **Function:** Fast identification of asylum seekers.				
UK Immigration Service	AFIS	Sagem	Live	–
Name of system: Portable 'QuickCheck' (MorphoTouch) from Sagem. **Function:** Fast identification of asylum seekers using GSM link from small ports.				
USA				
Boston Logan International Airport	Face	Identix (formerly Visionics)	–	–
Name of system: FaceIt technology from Identix (formerly Visionics), integrated by ATC Biometric Technologies. **Function:** Scan travellers' faces against a watchlist of suspected terrorists.				
Charlotte/Douglas International Airport	Iris	EyeTicket	Live	–
Name of system: EyeTicket Passenger Processing System. **Function:** Processes flight crew and employees to provide secure admission to the ramp and security bypass in the concourse.				

Table 3.10 *continued*

Name	Technology	Name of supplier	Status	No. enrolled
USA *continued*				
Continental Airlines	Fingerprint	BioScrypt	Live	–
Name of system: Control Module Inc's BioScan II product, integrated with Bioscrypt's MV1100 OEM solution with time clock functionality. **Function:** Secure personnel files and upgrade time and attendance systems.				
Dallas/Fort Worth International Airport	Face	–	–	–
Function: Surveillance.				
Federal Aviation Authority	Iris	Titan Corporation	–	–
Name of system: Secure Executive Toolset (SET) from Titan. This uses the Panasonic Authenticam iris recognition and video conferencing camera. **Function:** Access control to intranet site.				
Fresno International Airport	Face	–	–	–
Function: Surveillance.				
Immigration and Naturalisation Service (INS), US	Hand	EDS	Live	–
Name of system: INSPASS from EDS. **Function:** Automated passport control.				
Immigration and Naturalisation Service (INS), US	Fingerprint	Identix (formerly Visionics)	–	–
Name of system: CMS live scan fingerprint systems from Identix (formerly Visionics). **Function:** Applicant processing.				
Manchester Airport	Face	Viisage	–	–
Name of system: SecureFlight from Viisage. **Function:** Surveillance.				
Oakland International Airport	Face	–	–	–
Function: Matching faces of those brought into airport interrogation rooms against a database of known criminals.				
Palm Beach International Airport	Face	–	–	–
Function: Surveillance.				

Table 3.10 *continued*

Name	Technology	Name of supplier	Status	No. enrolled
USA *continued*				
Reagan National Airport	Fingerprint	Identix	–	–
Name of system: TouchPrint 2000 Applicant Fingerprint Systems from Identix. **Function:** Employee background checks.				
St Petersburg – Clearwater International Airport	Face	–	–	–
Function: Surveillance.				
United Airlines	Fingerprint	Identix	Live	70,000
Name of system: Live Scan system from Identix. **Function:** Employee access control.				

3.5.9 Commercial/Other

The market loosely covered by 'commercial/other' includes the growing use of biometrics within commerce (corporate access control, business-to-business applications, and business-to-consumer activities). Other areas of interest are in the domestic sector. Although this is currently not an area of huge growth, it does represent a significant market share, and provides an indication of where some future orders could originate. A large percentage of orders within this sector come from physical access control, network access control and time and attendance applications.

3.5.9.1 Summary of Market Applications

The largest aspect of this market is the use of biometrics within commerce:

- Corporate access control
- Corporate network access control
- Corporate time and attendance
- e-work
- Transportation and distribution
- Employee background checks
- Access control to high security – (non-governmental) – areas

Other areas include:

- Time and attendance systems for firefighters
- Automobile and truck security
- Residential access control
- Home PCs

3.5.9.2 Drivers

- *Cost savings* – The strongest driver within this market is the need to reduce fraud and cut costs. This can be seen through the use of time and attendance systems, for example, which eradicate 'buddy punching'.

- *Safety and security* – Within the commercial sector expect to see further demand for access control systems and employee background checks. A growth in interest from other areas of society is also expected in areas such as domestic access control and automobile security, for example.
- *Remote working* – The changing nature of business is proving a moderate driver within this market. As more people work remotely, approval processes within companies have had to be changed. Biometrics can be used to verify an approver's identity and can be stored in a searchable database to provide an audit trail of all transactions. We believe that this is a moderate driver in the short term because a lot of companies are still showing little interest in addressing some of the challenges of e-business. Equally, biometrics face stiff competition from other technologies, such as PKI systems.

3.5.9.3 Detractors

- *Competition* – The strongest detractor within this market is the huge amount of competition that biometrics face from other forms of authentication. This competition comes from traditional sources (keys, magstripe cards and PINs,) and new sources (smart cards, PKI).
- *Lack of knowledge* – Another detractor is the lack of public knowledge about biometrics. People still need to be informed about what biometrics are, how they work and what benefits they can offer. This is an important step if some of the more creative uses of the technology (e.g. ignitions in automobiles) are to become common place.
- *Cost* – The cost of biometrics is still too high for certain parts of the market. In particular, domestic PC users will need to see prices fall further if there is to be any significant take-up of the technology. This is a major detractor within the domestic market. In the commercial market, it is a moderate detractor in instances where a good business case can be made concerning cost-effectiveness.

3.5.9.4 Outlook

There are enormous possibilities for growth within the commercial sector. This is an area whose decision-making process is significantly faster than that of the healthcare, governmental and financial industries. It is an area that is embracing new technology throughout its hierarchy, and is – in times of economic growth – less cost conscious than any public sector organisation. Further interest from domestic users can also be expected to grow. As prices fall, and knowledge of the technology increases, a larger percentage of people may consider adopting the technology within their homes. We also expect further interest in the use of biometrics for more creative applications. Some of these may fail to go beyond the concept stage, but others could prove thriving markets in the long-term.

This market is substantial in revenue terms as it covers a multitude of industries, not individually large enough to warrant their own section. In dollar terms the sector was around US$43 million in 2000, but this is expected to grow to US$355 million by the end of the forecasting period.

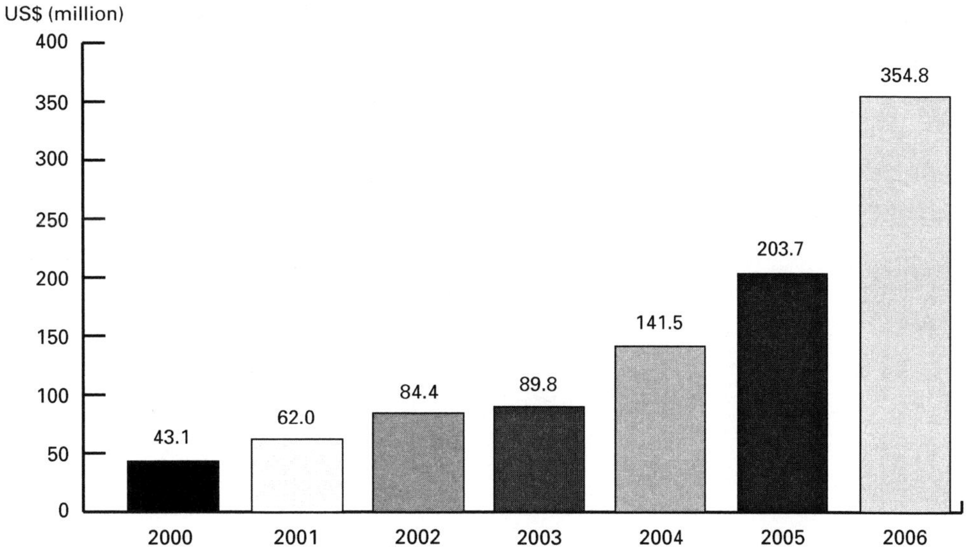

Figure 3.20 Revenue Forecast for the Commerical/Other Market 2000–2006 (US$ million)

3.5.9.5 A Selection of Commercial Market Installations

Table 3.11 A Selection of Commercial Market Installations

Name	Technology	Name of supplier	Status	No. enrolled
USA				
Kenworth Truck Company	Fingerprint	BioScrypt	Concept	–
Name of system: Uses fingerprint technology from BioScrypt. *Function:* Controls ignition to truck.				
The Manor, Yorktown	Speaker and face	Graphco and AcSys Biometrics	Live	–
Name of system: VoicePass from Graphco Technologies. Face Recognition from AcSys Biometrics. **Function:** Access control for residents. The technology is also used by residents in the local store allowing their accounts to be debited.				
Triangle Distribution Network	Fingerprint	Sense Technologies	Live	–
Name of system: Checkprint T/A clocking system from Sense Technologies. **Function:** Time and attendance of staff in strategic distribution centres.				
USA (group representing 9 nuclear utilities)	Fingerprint	Cross Match Technologies	Live	–
Name of system: ID 1000 live scan fingerprint systems. **Function:** Employee background checks for staff working in nuclear industry.				
West Hamilton Beach Fire Dept	Fingerprint	Sense Holdings	Live	–
Name of system: Checkprint T/A clocking system from Sense Technologies. **Function:** Time and attendance of fire fighters.				

3.6 Geographical Market Forecasts

This report breaks down the world into five regions – North America, Latin America, Europe, Asia Pacific and the Rest of the World.

Revenue figures are based on the location of the project geographically, not the location of the supplier – so a US-based company installing a system in Japan would apportion that revenue to Asia Pacific.

3.6.1 Outlook

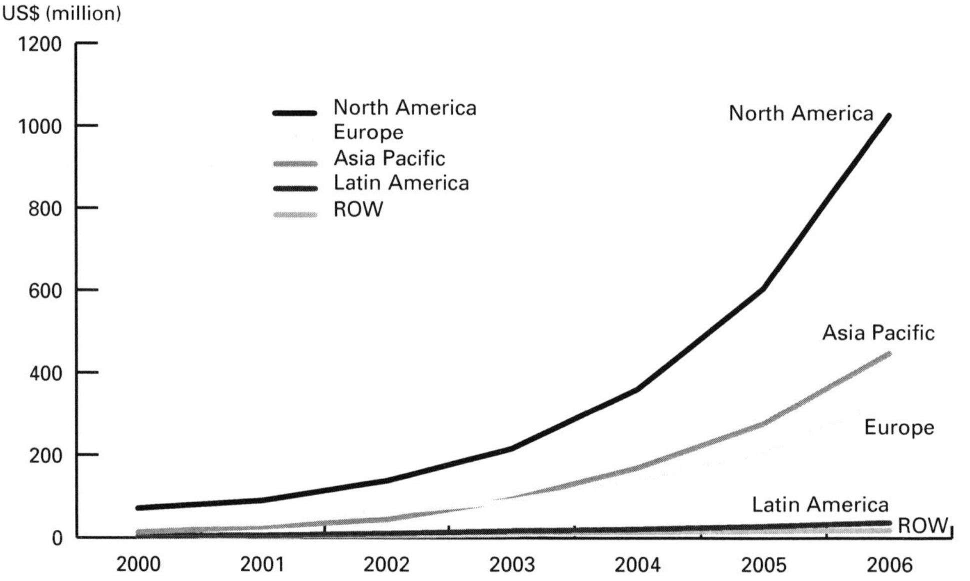

	2000	2001	2002	2003	2004	2005	2006
North America	71.0	89.9	137.9	216.0	360.7	605.4	1027.0
Europe	30.0	35.2	62.4	89.0	141.5	203.7	336.1
Asia Pacific	13.1	25.8	44.2	93.2	169.7	277.2	448.1
Latin America	4.2	5.8	10.4	16.9	21.2	28.3	37.3
ROW	1.6	3.7	5.2	8.5	14.1	17.0	18.7

Figure 3.21 Revenue Forecast for the World's Geographic Regions 2000–2006 (US$ million)

The US market still dominates the biometric industry and this is forecast to be the case until well beyond the forecasting period in this report. Major contracts from the US government and the aviation industry as well as the financial and healthcare sectors will underpin this revenue position. In particular the stimulus given to the biometric industry by the events of September 11th have been significant and primarily felt within the USA market.

Whereas the Asian Pacific region had been expected to rival the US market in our previous report, this is no longer the case. Instead Europe and Asia are pretty much level pegging in terms of revenue until 2004, at which time there is a pushing forward by Asian markets to overtake the slight European advantage.

Latin America is expected to make steady progress over the next four years, with most applications expected to come under the government and law and order markets.

The Rest of the World category in the report includes the Middle East and Africa. There have been some notable projects in both these regions, but in revenue terms the overall industry contribution is expected to remain small.

More specifically, African nations are tending to use biometrics in the area of national ID and voting control as the countries put in place solid social infrastructures designed to stamp out fraud.

Middle Eastern countries, meanwhile, are using biometrics for a diverse number of reasons. The biggest market in this region is Israel, which has seen immigration control installations, as well as some notable financial industry case histories, such as the use of speech verification technology at First Direct Bank and dsv technology at Bank Hapoalim.

4 Company Profiles

This chapter provides in-depth information on a selection of the industry's leading biometric companies, as well as some companies that are interesting from a strategic standpoint. The list, organised by technology grouping, is displayed below:

- **Dynamic Signature Verification**
  Communication Intelligence Corporation (CIC)
  Cyber-SIGN Inc
- **Face**
  Viisage Technology Inc
  Visionics Corporation (see Identix)
- **Fingerprint**
  AuthenTec Inc
  Bioscrypt Inc
  Infineon Technologies AG
  Precise Biometrics AB
  SecuGen Corporation
  STMicroelectronics
- **Hand Geometry**
  Recognition Systems Inc
- **Iris Recognition**
  Iridian Technologies Inc
- **Middleware**
  BioNetrix Systems Corporation
  I/O Software Inc
  Keyware Technologies
- **Speaker Verification**
  Buytel (now Voicevault)
  Nuance Communications Inc
  Persay Inc

4.1 AuthenTec Inc

PO Box 2719
Melbourne
FL 32902-2719
USA
Tel: +1 321 308 1300
Fax: +1 321 308 1430
Email:
mkt@authentec.com
Web:
www.authentec.com
President & CEO:
Scott Moody

4.1.1 History

AuthenTec Inc was founded in October 1998. The initial focus of the company was the creation and development of its extensively patented TruePrint Technology and the production of its first fingerprint sensor product line – the FingerLoc AF-S1. AuthenTec commenced as a spin off from Harris Corporation of Melbourne, Florida, which is one of the USA's largest international communications equipment companies.

In May 2000 the company introduced the FingerLoc AF-S2, an evolutionary successor to the original sensor. Also in May that year AuthenTec announced the world's first self-contained BIOS-level pre-boot authentication system – a product that can reliably prevent unauthorised use of any modern computer.

In September 2000 the company announced the release of the EntréPad AES4000 sensor, a low-profile, low-power product aimed directly at the PC and personal electronic device market.

4.1.2 Profile

AuthenTec is a fabless semiconductor company, based in the USA. It provides advanced silicon solutions to the biometric and security markets with applications in personal computers, wireless communications, access control and personal electronic devices. All of AuthenTec's products are based on the company's patented TruePrint technology, which reads below the surface layer of the skin to the 'live layer' of the fingerprint.

The company has a range of sensors, which use special software techniques to allow the sensor area to be smaller than many other manufacturers' touch sensors. This allows AuthenTec to compete well on price, as the amount of silicon used in the sensor is directly proportional to the price of the product. In addition AuthenTec attempts to use standard components where possible, such as sensor coating, in an attempt to keep costs low.

AuthenTec has raised monies from an impressive list of venture capital firms throughout the world. The company presently has been issued with or filed more than 30 patents.

4.1.3 Technology

As mentioned above, all of AuthenTec's products are based on the company's patented TruePrint technology. This is designed to achieve maximum fingerprint acquisition. Additionally the company's products perform dynamic optimisation on up to 15 parameters (such as old/young, dry/greasy), which allows the adjustment of the sensor to obtain the best image.

On the company's smaller sensors a software technique called Mosaic is used. At enrolment, the software effectively stitches together the smaller fingerprint images that are captured in order to form a larger overall fingerprint. Following enrolment the overall image size is around 90 mm^2. After around 30 touches under normal operation the size of the resultant template increases to an effective area of 169 mm^2, the company claims.

The company has two families of sensor – the FingerLoc (for harsh environments) and the EntrePad (for PCs, mobiles, PDSa etc). The FingerLoc AF-S2 has an active area of 13 × 13 mm, while the FingerLoc AFS8500 has an active area of 95.1 mm^2. The company's EntrePAD range comprises the AES4000 with an active area of 95.1 mm^2 and the AES3500 at 42.25 mm^2.

4.1.4 Strategic Information

Sales of PC equipment in the world have slowed over the last few years and AuthenTec is likely to be feeling the pinch from the repercussions. In order to bolster sales it has a range of sales representatives and distributors around the world. The company recently added numerous representatives in the European and Asian markets.

An important and prestigious customer for AuthenTec in recent months was IBM, which announced that its ThinkPad portable computer range would have the option to add fingerprint security using the AuthenTec's sensors.

From a technical standpoint the company's strategic position is clear – cheap sensors with good image acquisition properties. Some manufacturers are producing swipe-based sensors, where a person sweeps their finger across a very thin strip of silicon to give their fingerprint. Despite the cost advantages and small footprint of such a sensor, AuthenTec has steered away from this technique, as it believes the touch sensors will prove more acceptable to consumers.

4.1.5 Financial Information

Authentec is a privately owned company and does not publicly release revenue statistics. However elsewhere in the financial arena, the company recently secured US$20 million in Series B financing led by a US$5 million investment by Texas Instruments Inc (TI). All of AuthenTec's previous investors reportedly participated in this second financing round, with major contributions by Harris Venture First, Knickerbocker LLC, Advantage Capital and the China Development Bank.

This new series brings the total investment of venture based financing in AuthenTec to over US$35 million since its founding in October 1998.

4.2 BioNetrix Systems Corporation

1953 Gallows Road,
Suite 500
Vienna
VA 22182
USA
Tel: +1 703 734 9200
Fax: +1 703 734 6553
Email:
info@bionetrix.com
Web:
www.bionetrix.com
CEO: Steve Walker

4.2.1 History

BioNetrix Systems Corporation is a privately held company and is headquartered in Vienna, Virginia, in the USA. The company was founded in 1997.

4.2.2 Profile

BioNetrix develops middleware software for centrally managing user authenti-ation in enterprise and internet security applications. The BioNetrix platform unifies authentication management, providing integrated support for more than 20 advanced authentication methods, including smart cards, tokens and biometrics.

4.2.3 Technology

The company's main product is the *BioNetrix Authentication Suite 4.1.* This software platform allows organisations to deploy any combination of biometric (fingerprint, voice, face, iris and signature) and non-biometric (token, smart card and password) user verification technologies and operates seamlessly with other security applications, such as public key infrastructure (PKI), virtual private networks (VPN) and single sign-on (SSO).

4.2.4 Strategic Information

BioNetrix has developed strategic partnerships with numerous authentication technology companies, application software providers, digital certificate and encryption companies, value-added resellers and system integrators. These partnerships ensure that the BioNetrix product is deployed along with leading enterprise and web-based applications, as well as other network security solutions.

Its partners include many members of the biometric industry including AuthenTec, Bioscrypt, Identix, Polaroid, SecuGen, SONY, Veridicom, Iridian Technologies, Communication Intelligence Corporation, Cyber-SIGN, Nuance and Veritel.

The company sees the healthcare sector as one of special significance. In particular HIPAA legislation in the USA is proving to be a key driver for its business and it has already secured a number of contracts in this area, including the Lee Moffitt Cancer Center in Tampa, Florida, and St Luke's Episcopal Health System in Houston, Texas.

The company also sees the development of application single sign on as an important future requirement, and has recently added this functionality to its Authentication Suite. This now means that users will no longer have to have multiple application-specific passwords, as these are replaced with one single authentication method, such as a biometric. The applications that a person can access are controlled centrally.

4.2.5 Financial Information

The company does not disclose its revenue figures.

4.3 Bioscrypt Inc

Corporate & Canadian Office
5450 Explorer Drive,
Suite 500
Mississauga
Ontario
Canada L4W 5M1
Tel: +1 905 624 7700
Fax: +1 905 624 7742
Email: jean-pierre.lee@bioscrypt.com
Web: www.bioscrypt.com
President & CEO:
Pierre Donaldson

US Office
5805 Sepulveda Blvd,
Suite 750
Van Nuys
CA 91411
USA
Tel: +1 818 304 7150
Fax: +1 818 461 0843
Email: Jason.Nilles@bioscrypt.com

4.3.1 History

Bioscrypt Inc was formed in 2001, when two established companies – Mytec Technologies and Biometric Identification Inc (BII) – joined forces in a deal worth approximately US$25 million. Mytec was a Canadian company with an expertise in wireless biometric applications, while US-based BII was rooted in more traditional markets, such as physical access control and time/attendance applications.

The merger therefore created a company with a good strategic fit, which would be able to experience immediate success in the access control markets, but with a strong future growth opportunity in the wireless markets. The company owned and had rights to 15 biometric patents.

4.3.2 Profile

Bioscrypt is today one of the leading biometric companies in the fingerprint sector. The company has deployed over 35,000 readers for physical access and time and attendance, and has licensed its algorithm to enable over 65,000 devices for network access. The company's algorithm is interoperable with most of the leading fingerprint sensors and is both platform and operating system independent.

The company brands its technology as "bioscrypt on board", offering packaged products, OEM components and software licensing to security solution manufacturers and integrators worldwide for physical, wireless and network security applications.

4.3.3 Technology

Bioscrypt focuses on fingerprint recognition, using its patented Ridge Recognition algorithm. Pattern recognition methods are also suitable for use with other biometric types such as iris and face, and many of Bioscrypt's patents provide patent coverage for a broad range of biometric types. Unlike traditional minutia based algorithms, Bioscrypt's pattern recognition algorithm uses the entire fingerprint image to authenticate the user, rather than selected points on the print. According to the supplier, this makes it more robust to distortions, such as cuts or scratches and other day-to-day variations of the finger.

At the heart of all the company's products is its verification library Bioscrypt Core. The company's product range comprises the following solutions:

- *V-Smart* – This is a two-in-one solution that combines a biometric device and smart card reader for access control. It uses standard MIFARE contactless smart card technology and allows for Wiegand communication formats up to 64-bits. There is also an iCLASS version that allows support for smart cards from HID.

- *V-flex* – The V-flex makes it possible to add the security of fingerprint verification to any reader that is capable of providing Wiegand communication in an industry standard format. In other words organisations with keypads, bar code readers and proximity card readers are able to add fingerprint verification to their existing systems without replacing the existing reader infrastructure.
- *V-prox* – The V-prox combines a self-contained fingerprint verification package about the size of a business card with a proximity card reader.
- *V-Pass* – The V-Pass eliminates the need for pins, keys or cards to gain access to secure areas. To gain access, the user places their finger on the reader and the V-Pass confirms or denies access based on a database of fingerprints from authorised users. The V-Pass performs enrolment, identification and template storage for 200 fingerprints.
- *V-Station* – The V-Station supports standard and proprietary Wiegand communication formats and stores up to 3,000 templates on the unit. It has an optional integrated proximity or contactless smart card reader.
- *MV1200* – The MV1200 is a full-featured stand-alone fingerprint verification module. The device is smaller than a business card and provides an independent and flexible finger mask with an external mounting area for capturing an image the size of a postage stamp.

4.3.4 Strategic Information

Bioscrypt is focused on selling its biometric technologies in the areas of physical access, network access and wireless access through a combination of products and technology licensing. The company works locally and globally with a range of business partners including resellers, distributors, systems integrators and solutions providers.

Its biometric technology is deployed in such markets as healthcare, government, defence, automotive and financial services. Among the numerous companies and partners using the company's technology are the US Army, NASA, Continental Airlines, Intel, Atmel, HID Corporation, Honeywell and Northern Computers.

Bioscrypt believes that the secure deployment of biometric solutions is on the verge of rapid expansion, and that the company's technologies are well positioned to take advantage of this emerging market.

Recently the company received a strong boost to its technology ratings, as it performed excellently in the international Fingerprint Verification Competition 2002, an independent evaluation of fingerprint matching technologies. This test pitted the company's algorithm against 30 other companies and Bioscrypt won 19 of the 24 categories.

4.3.5 Financial Information

Bioscrypt has been listed on The Toronto Stock Exchange since 1994. As of 1 May 2001 its shares traded under the symbol "BYT".

Bioscrypt Inc

Key Figures (C$ million) Year ended 31.12

	2001	2000
Revenues	8.4	0.3
Profit/(Loss)	(24.8)	(17.7)

The company's revenue is primarily derived within the Americas, which accounts for approximately 75% of worldwide sales. Currently, the most significant portion of revenue is from the sale of biometrics readers used in physical access control applications. These units are manufactured in Asia under contract to Bioscrypt, and are sold through a distribution channel of security company dealers, value-added resellers and systems integrators. Historically, profit margins on these products have been approximately 15–20%, however, the company has recently enhanced certain components within these products that it believes will improve margins.

The second source of revenue is from licensing the company's software algorithms to either manufacturers of biometric sensors, who in turn sell to manufacturers of wireless and network access devices (cell phones, personal digital assistants, laptop computers, etc.), or directly to these device manufacturers. Revenue from licensing agreements consists of an up-front technology transfer fee, a perpetual per unit royalty and an annual maintenance fee for support and software upgrades.

In some instances the company will receive revenue for consulting and non-recurring engineering services. Margins for license revenue are approximately 95%, and represent the largest growth opportunity for Bioscrypt.

4.4 Communication Intelligence Corporation (CIC)

US Headquarters:
275 Shoreline Drive,
Suite 500
Redwood Shores
CA 94065-1413
USA
Tel: +1 650 802 7888
Fax: +1 650 802 7777
Email: sales@cic.com
Web: www.cic.com
Chairman, President & CEO: Guido DiGregorio

China Headquarters:
CICC Ltd 18th Floor
285 North Zhongshun Rd
Nanjing, PR 210003
China
Tel: +86 25 3341046
Fax: +86 25 3341043

4.4.1 History

Communication Intelligence Corporation (CIC) was incorporated in the state of California in 1981 and reincorporated in the state of Delaware in Oct 1986, as a wholly owned subsidiary of a predecessor corporation with the same name. The company has operated a 90%-owned joint venture, Communication Intelligence Computer Corporation, with the Jiangsu Hongtu Electronic Group, in Nanjing, China, since September 1993. In October 2000, CIC acquired certain intellectual property of PenOp Limited in exchange for 4.7 million shares of common stock of the company.

4.4.2 Profile

CIC is a supplier of electronic signature, biometric security and natural input solutions focused on emerging applications such as paperless workflow, smart wireless devices and e-commerce. The company is one of the dominant players in the signature verification space. The company's core software technologies include multilingual handwriting recognition systems, biometric signature verification, natural messaging, and operating system extensions that enable pen input.

CIC sells direct to OEMs and enterprises. Products are also available through major retail outlets such as Circuit City, CompUSA, Staples, OfficeMax, as well as key integration partners and direct via its own website. Many large companies such as Ericsson, Fujitsu, Hitachi, Siebel Systems, IBM, Charles Schwab and Prudential have licensed the company's technology, but not all have chosen to use the supplier's biometric offering.

4.4.3 Technology

CIC's software does not require specialised hardware and works with virtually all PDA devices, touch-sensitive screens and digitising tablets. Within its biometric product range CIC has four main products:

- *Ink Tools* – A developer's resource for creating the core components needed to capture, secure and biometrically validate signatures. It is used to integrate signature technology with existing systems. It was used to create the first version of Sign-it (see below).
- *Sign-On* – Provides log-on security to handheld devices. A person enrols in the system by signing their name a number of times to create a suitable template, which is captured, verified and encrypted directly onto the device. Subsequently, a user signs their name to unlock the device and gain access to data on their organiser.
- *Sign-it* – Enables individuals to electronically sign documents. Once signed, the biometric-based signature is bound to the document which can then be securely transferred electronically. This product enables users to complete business transactions requiring a signature online, as well as enabling the elimination of paper.
- *iSign* – Provides internet-based signature verification for use in network authentication or electronic document signing applications. Specifically designed for the Web environment, it provides an easy-to-use interface for anyone developing content in Windows or Java to integrate biometric signatures as part of their application.

4.4.4 Strategic Information

The company's products are marketed through three main avenues: OEMs, enterprise and online/retail sales. OEM sales efforts are aimed at license revenues derived from primarily smart handheld device manufacturers. Enterprise sales efforts are directed at both software providers and end-users. Online sales represent revenues generated from the company's software sold via its website, www.cic.com.

CIC has many large, fully deployed customers in the United States, Europe and China, with at least 74 to date in the financial services space. Most of these do not actively use the biometric capabilities of CIC's product range, although some customers such as Charles Schwab, Prudential, American General Life (AGLA), Nationwide Building Society (UK) and Agricultural Bank of China are active users.

The supplier believes that legislation legalising electronic signatures in most countries of the world will provide a major boost to its operations as will an increased move toward the paperless office. September 11th, 2001 generated many lines of enquiry for the company.

In the USA, CIC is the dominant player in the signature verification space with competition from few other players in the sector. There are pockets of competition overseas, however, such as WonderNet in the Israeli market and Cyber SIGN and Softpro in Europe and Asia. CIC believes that some of these players are infringing its patents and warns it is taking this issue extremely seriously.

4.4.5 Financial Information

For the five-year period ended December 2001, operating losses aggregated approximately US$21 million and at the end of 2001, the company's accumulated deficit was approximately US$76 million. No individual biometric revenue segmentation was available. As of March 2002 there were approximately 91 million shares of CIC common stock outstanding.

CIC

Key Figures (US$ 000) Year ended 31.12

	2001	2000	1999	1998
Revenue	5,947	7,312	6,518	4,581
Net Losses	(2,946)	(1,607)	(1,722)	(3,285)

4.5 Cyber-SIGN Inc

180 Montgomery St,
Suite 925
San Francisco
CA 94104
USA
Tel: +1 408 324 1001
Fax: +1 408 324 1057
Email:
info@cybersign.com
Web:
www.cybersign.com

4.5.1 History

Cyber-SIGN Inc is predominantly a Japanese company with a wholly owned subsidiary in Europe. The company was originally formed in December 1999 through a joint venture involving a number of players including Cadix (the original technology developers), Fujitsu and Nikko Capital. Cyber-SIGN also had an operation in the USA, which was wholly owned by Cadix. This was shut down in 2001, however, as margins in the operation were not good enough. At this point Cyber-SIGN in Japan took control of the shell company.

Due to weak economies in both the USA and Japan over the last year, the company in the USA remained dormant, but has now reportedly been given a new lease of life. It is expected that either a new company will be launched or a revitalised Cyber-SIGN with new management will be announced.

4.5.2 Profile

Cyber-SIGN operates in the area of biometric signature verification, signature capture and display. Cyber-SIGN is the supplier's client/server signature verification development tool. The company's expressed strategy is to establish its electronic signature technology as a de-facto standard.

The bulk of the company's business still occurs in Asia, where sales are claimed to be moving ahead rapidly. Since September 11th, however, the company has witnessed increased interest from the USA and is now focusing more effort on this market by opening a new office in the USA. According to the company this was necessary as it was finding it tough to be globally successful without a US operation.

4.5.3 Technology

Cyber-SIGN software looks at the behavioural patterns of a hand written signature to confirm the identity of a computer user. The company's products centre around four dimensional dynamic signature verification technology which measures variables, such as changes in timing, velocity and pressure. Similar to most other signature verification companies, its technology is relatively cheap, requiring no special hardware except for a digitising tablet to capture signatures for verification.

The company's main product range includes *Cyber-SIGN Enterprise*, a client server technology, and *Cyber-SIGN Personal* a local authentication design for non-networked user authentication. There are also application and plug-in products designed to operate with packages such as Adobe Acrobat and Microsoft Windows, as well as a screen saver product.

4.5.4 Strategic Information

Along with most other digital signature verification (dsv) companies, Cyber-SIGN believes that the signing of the Electronic Signatures Global and National Commerce (E-Sign) Act by the US Federal Government, and similar legislation internationally, will boost the company's sales. Potential applications for the company's products include electronic commerce, finance, healthcare, on-line banking, government, legal, security, workflow and other markets that require non-refutable, tamper-proof documents.

Cyber-SIGN is already proving itself successful in Asian markets – in particular at Japanese telecoms company NTT, where some 100,000 employees are now believed to be using its dsv technology to secure email and to connect to the corporation's central server for printing and signing letters from their desktop PCs. Whether the company can secure a foothold in the US market in the face of competition from companies such as CIC remains to be seen. However, the company's long-term goal is to secure venture capital and float on the US stock exchange.

The company's main competitors include CIC in the USA, Softpro in Europe and Asia and to a lesser extent Wondernet, which operates primarily in the Middle East.

4.5.5 Financial Information

The company does not publicly disclose its revenue statistics. Sales are estimated at more than US$3.5 million in 2001.

4.6 Identix Inc

5600 Rowland Road
Minnetonka
MN 55343
USA
Tel: +1 952 932 0888
Fax: +1 952 932 7181
Email:
sales@identix.com
Web: www.identix.com
President & CEO:
Joseph J Atick PhD

4.6.1 History

Founded in 1982, Identix Inc became a public company in 1985 and traded on the American Stock Exchange. The company pioneered the development of patented live-scan fingerprint systems and biometric identification systems. Significant milestones include:

- Acquired Identicator Technology in April 1999
- Acquired Identix Public Sector (formerly Anadac Inc) in 1992
- Visionics Corporation formed in 1994
- Visionics Corporation and Digital Biometrics merge to form the publicly traded Visionics Corporation in February 2001
- Identix and Visionics merged in June 2002

4.6.2 Profile

Nasdaq-listed Identix recently merged with Visionics Corporation and is now possibly the world's leading multi-biometric security technology company. The company has a broad range of fingerprint and facial recognition technology offerings that serve a range of industries and market segments – most notably, government and law enforcement, aviation, financial, healthcare and corporate enterprise. In addition, the company's wholly owned Identix Public Sector (IPS) division provides project management and facilities engineering services to the government sector.

Identix was an industry leader in fingerprint identification. Visionics, meanwhile, was a significant player in the facial recognition market (and the livescan market following its merger with Digital Biometric Inc.). By joining forces, Identix has become the biggest multi-biometric biometric company and is actively capitalising on synergies created.

4.6.3 Technology

Identix's technology is diverse, ranging from fingerprint scanners and software to facial recognition systems and livescan devices. The company organises its product range into a number of categories – security platforms, information security, access control, facial surveillance, livescan, mobile identification and embedded solutions.

The core biometric products in its range include:

- *The DFR200* – This is an optical fingerprint sensor and is incorporated into many OEM products.
- *Biologon software* – The company's PC and network security software for use with fingerprint-based biometric accessories. Available for Windows, Novell and Wyse platforms.
- *itrust* – A biometric authentication management platform.

- *Biometric Network Platform (BNP)* – A scalable technology framework for applying biometric technologies in large enterprises or over multi-camera networks.
- *FaceIt ARGUS* – A full-time facial surveillance and identification engine that allows each camera, whether new or old, to serve as a diligent observation point, even when the video is not being actively observed. FaceIt can also be used on its own for IT security type applications.
- *Fingerscan V20* – Physical access control offering.

There are many other products, but not all fit into the strict category of biometric devices, such as the livescan devices. Other biometric products include embedded solution SDKs and software, such as its Fingerprint Bio-Engine.

4.6.4 Strategic Information

Identix has developed a new organisational structure that is more customer orientated driven and cost efficient. Instead of being focused around product lines as in the past, Identix has established four specific business groups, each focused on specific markets and customers – federal, state/local government, commercial and international. The strategic groups will have focused sales and business development teams that sell Identix's full range of biometric products and solutions, which will be supported by shared fulfilment and functional groups that deliver the engineering and operational requirements for the company.

Generally speaking, customers are encouraged to adopt Identix products and technologies through system integrator, OEM, distributor and reseller partners, although Identix may provide its products and services directly to the customer.

Identix recently shocked the biometric market by pulling substantially out of the low cost fingerprint sensor market, where it had been one of the market leaders. In particular it has halted development of the DFR400 optical fingerprint sensor market. The company can make greater margins selling software than its sensor hardware. Originally the company sold its DFR sensor range to accompany its software. The company now says that silicon-based sensors may be more appropriate as they have become smaller, more stable and cost effective in recent years.

Identix sees itself as the market leader and recently pleaded for companies to work with it, rather than against it. Whether it is too early for the industry to dance to this tune is difficult to tell, but the merger has certainly made the company a dominant force in the industry. The company has faced criticism by its peers for overstating the capability of its systems, especially in the facial recognition arena. While this may be true to a certain extent, especially in the immediate aftermath of September 11th, the company is finding success in this area. It answered its critics recently by performing well in airport trials and at Virginia Beach, where its surveillance technology had to pass defined acceptance criteria.

4.6.5 Financial Information

Identix has not paid any cash dividends in the past and does not expect to in the foreseeable future. It is too early to tell whether the merger will deliver all that it has promised, but early indications are that the company is well positioned now to benefit from any major upturn in the market.

In terms of revenue, at the time of writing this report, there were no clear figures as to how the combined results will look. The following is a look at the revenue of Visionics and Identix separately, before the merger.

Visionics Corporation

Key Figures (US$ 000) Year ended 30.9

	2001	2000	1999
Revenues	30,473	25,575	24,822
Of Which:			
FaceIt Licenses	2,978	1,760	1,473
FaceIt Services	1,487	875	1,150

Identix Inc

Key Figures (US$ 000) Year ended 30.6

	2001	2000	1999
Revenues	81,828	72,695	81,762
Of Which:			
Imaging	27,010	22,313	27,170
Security	8,051	8,612	7,058
Government Services	50,561	45,463	50,265
Elimination of Inter-segment Revenues	(3,794)	(3,693)	(2,731)

4.7 Infineon Technologies AG

St-Martin-Str 53
81669 Munich
Germany
Tel: +49 89 234 00
Fax: +49 89 234 2 46 94
Email: company.info@infineon.com
Web: www.infineon.com
President & CEO:
Dr Ulrich Schumacher

4.7.1 History

In April 1999 Siemens Semiconductors became Infineon Technologies AG. Siemens had vast knowledge in security technology and applications, communication technology, chip card technology and more. In the biometrics arena Siemens started with fingerprint recognition for police identification systems in the early 1990s. Siemens had a long history in the development and evaluation of biometric systems, such as voice or face recognition, fingerprinting and signature verification. Combining this experience and knowledge with advances in technology led to the development of the CMOS-based FingerTIP.

This expertise has now passed to Infineon since its divergence from the parent group.

4.7.2 Profile

Far from being focused on biometrics, Infineon Technologies, based in Munich, Germany, offers a multitude of semiconductor and system solutions for applications in the wired and wireless communications markets, for security systems and smartcards, for the automotive and industrial sectors, as well as memory products.

With a global presence and approximately 29,000 employees, Infineon is one of the market leaders for chip card ICs, and has extensive skills in many areas of security technology. This expertise has been extended into security ICs for PCs and electronic commerce markets. Infineon manufactures the FingerTIP biometric sensor, for the evaluation of fingerprints as a means of convenient and secure identification. Infineon in collaboration with partners in its Silicon Trust programme, offers complete security solutions.

4.7.3 Technology

FingerTIP uses the capacitive difference between the surface of the sensor and the ridges and valleys of the finger. The capacitance measured at a ridge differs from that at a valley. Thus, about 65,000 capacitors in a field of 224 × 288 pixels acquire and transform the data into a Bitmap image. This signal is subsequently passed to the recognition algorithms.

The FingerTIP sensor has an active sensor area of 11 × 14.3 mm and provides a dpi resolution of 515. The company claims to have been the first to provide ESD protection of ±15 kV and over.

The supplier's system provides encoder and matcher biometric algorithms in software or firmware and has a BioAPI compliant SDK. The company is working on producing an ISO standard smart card solution, which could be used in a credit card scenario.

4.7.4 Strategic Information

The company is believed to have experienced rapid growth in the sales of its sensors since 1999. It sells the bulk of its sensors in the computing markets, although smaller percentages are also used for physical access control and financial transactions. Based in Europe, the company does best in this market, but also sells a significant proportion of its sensors in Asia and North America.

The company warns of a general market stockpiling of fingerprint sensors in the market at present, which could curb future sales in the short term. This glut of sensors has been caused by many different factors, but one significant factor was September 11th, which caused many computer peripheral manufacturers to order greater quantities of the devices. This would have been fine, but for the general IT security market slowdown, which caused companies to hesitate putting money into upgrading their infrastructures.

The company's immediate competitors in the silicon sensor arena include Atmel, AuthenTec, Fingerprint Cards, Fujitsu, STMicroelectronics and Veridicom.

4.7.5 Financial Information

There is no specific biometric industry segmentation in Infineon's public figures, but in the fiscal year ended September 2001 the company achieved total (all group) sales of €5.67 billion. Security & Chip Card IC revenue amounted to €588 million.

Infineon is listed on the DAX index of the Frankfurt Stock Exchange and on the New York Sock Exchange.

4.8 I/O Software Inc

1533 Spruce Street
Riverside
CA 92507
USA
Tel: +1 909 222 7600
Fax: +1 909 222 7601
Email:
info@iosoftware.com
Web:
www.iosoftware.com
CEO: William Saito

4.8.1 History

Founded in 1991, I/O Software Inc (IOS) is one of the leading application and system security software providers supporting numerous types of user authentication including biometrics, Public Key Infrastructure (PKI) and smart cards. IOS is also the only company whose user authentication security software infrastructure will be built into the next generation of Microsoft's Windows operating system following an agreement with Microsoft in May 2000.

4.8.2 Profile

I/O Software is a developer of flexible information security software infrastructure for user authentication, encryption, and Internet/e-commerce using PKI and digital certificates. I/O Software's middleware solution supports a broad range of authentication technologies including biometrics (fingerprint, iris, retinal, voice and facial recognition), smart cards, RFIDs and tokens. I/O Software's strategic technology partners include Microsoft, Intel, Sony, AuthenTec and Iridian.

4.8.3 Technology

I/O Software's flagship product is SecureSuite. It supports multiple combinations of authentication technologies on a single network and is offered in nine different languages. The administrator can assign one or any combination of different authentication methods to a user including fingerprint, password, smart card, RFID, token and iris. Since SecureSuite provides a consistent user interface for widely varying authentication technologies, users can change from one technology to another without re-training. SecureSuite is compatible with Windows 95/98, ME, NT 4.0, NT 4.0 Server, 2000, 2000 Server and XP.

IOS has made public key cryptography an integral part of its software infrastructure and has laid the groundwork for interoperability with external Public Key Infrastructure (PKI) technologies. The company believes that PKI along with biometric and/or token-based authentication technologies will be the key to widespread adoption of advanced information security software.

All of IOS's applications are based on BAPI technology.

4.8.4 Strategic Information

IOS has established working relationships with many key players in the authentication software and hardware industries including Sony, Microsoft, Intel, Visa, Veridicom, Toshiba, Fujitsu, NEC, Mitsubishi, Iridian Technologies, Oki, Visionics (now Identix), SecuGen, AuthenTec, Rainbow Technologies, Kingston, Panasonic and Targus.

IOS has also established partnerships with independent hardware vendors (IHVs), value added resellers (VARs), original equipment manufacturers (OEMs), independent software vendors (ISVs) and system integrators (SIs).

In May 2000, IOS and Microsoft signed an agreement to integrate IOS's Biometric Application Program Interface (BAPI) and other core technology into future versions of the Windows operating system. Furthermore, in December 2000, IOS and Intel signed an agreement to use IOS's software to authenticate a user before proceeding to boot the system. As a result of these agreements, IOS is in a unique position as it is able to provide information security software, services and solutions that are forward compatible with Microsoft's future product lines from boot-up to network logon to application access.

It is believed that Microsoft has completely reworked the BAPI technology. Suppliers conforming to the more widespread BioAPI standard will hope an appropriate translation layer can be established as soon as possible. Until then I/O Software is clearly in a strong short-term market position.

4.8.5 Financial Information

As a private company IOS does not release financial data publicly.

4.9 Iridian Technologies Inc

Main Headquarters
121 Whittendale Drive,
Suite B
Moorestown NJ 08057
USA
Tel: +1 856 222 9090
Email:
iridian@iridiantech.com
Web:
www.iridiantech.com
President & CEO:
Bill Voltmer

Iridian SA (Geneva)
Lé Tresi 6C, 1028
Préverenges,
Switzerland
Tel: +41 21 804 8866

4.9.1 History

IriScan was incorporated as a Delaware Corporation in 1990. In 2000 IriScan joined forces with licensee Sensar Corp and the company changed its name to Iridian Technologies Inc. Today the company employs 25 people and has offices in Geneva and Moorestown, New Jersey.

4.9.2 Profile

Iridian Technologies owns a fairly impregnable set of patents for iris recognition technology. In particular Iridian's US patent 4,641,349 entitled "Iris recognition system" details the use of the iris as a means of biometric identification. This makes the company the main player in this technology space and as such it leads the world in research, development and marketing of authentication technologies based on iris recognition. (Note that Senex Technologies in Korea recently launched an iris recognition device, which Iridian said infringes its patents.)

The company is primarily a supplier of biometric software with investors and Fortune 1000 customers in banking, transportation, healthcare and security as well as law enforcement and government.

Establishing a global presence, Iridian Technologies Inc created a wholly owned European subsidiary, Iridian Technologies SA, in Switzerland. This entity leads the marketing, sales and customer service efforts throughout Europe, the Middle East and Africa (EMEA).

4.9.3 Technology

Iris recognition is based upon the extremely unique pattern of the eye's iris. A number of objective tests and evaluations over the last eight years have identified iris recognition technology as the most accurate biometric. The most recent of these evaluations was reported by the UK's National Physical Laboratory in April 2001.

Iridian's products include:

- *IrisAccess* – Developed in partnership with LG Electronics, this product is designed for physical access security. Iridian's iris recognition technology software is used as part of the solution.
- *Panasonic Authenticam* – Enabled with Private ID iris recognition technology from Iridian Technologies this handheld device is designed for use in information access solutions. It includes I/O Software's SecureSuite software to allow multiple users to securely access PCs, files, folders, applications and password banks. The camera can also be used for video conferencing and online collaboration.
- *Private ID software* – Allows an iris recognition camera to capture, select, and secure iris images. These images can then be processed by the KnoWho Authentication Server. The server will check that the images are secure before doing biometric operations, such as enrolment or recognition.

4.9.4 Strategic Information

Iridian Technologies has recently completed the transition from a hardware and software manufacturing company to a software vendor. As a result of this shift, the company has streamlined its operating expenses and it no longer has a large, direct sales force.

Iridian Technologies works closely with authorised VARs (value added resellers), systems integrators, catalogue vendors, and hardware and software resellers who customise solutions for specific markets. The iris recognition product portfolio is expanding with new products that will support large database and public use applications. Both Panasonic and Oki have released public use, walk-up Private ID-enabled imagers, which should be available in late 2002.

Iridian's new focus on the Transportation and Government markets is actively shaping how it aligns its products, organisation, messaging and pricing. Iridian says it is fully committed to this initiative and is working to ensure that it can meet the challenges of delivering public use applications such as simplified passenger travel, restricted access and border control.

Consistent with its focus on the transportation and government markets, Iridian launched a targeted marketing campaign from July through October 2002 via various aviation related magazines.

Notable companies using the supplier's technology include:

- **Transportation (Border Control, Simplified Passenger Travel, Restricted Access)** – Amsterdam Schiphol Airport, Heathrow Airport, Saudi Arabia Border, Singapore/Malaysia Border.
- **Government** – US House of Representatives, Brookhaven National Labs, NSA, Department of Defense, Federal Aviation Authority, Pentagon Officers Athletic Club, Office of the Chief of Staff Army.
- **Law Enforcement** – Sarasota County, FL; Lancaster County, PA; York County, PA; United Arab Emirates Prisons.
- **Healthcare** – Eagleville Hospital, PA; University of South Alabama's Hospitals including Womens and Childrens, Knollwood and Medical Center.
- **Entry Access for Utilities, Data Centres, etc** – Ameren Union Nuclear Power Plant, Verisign Data Centers, Quantas Data Center, Oracle Data Centers, Columbian Army and Secret Service.

4.9.5 Financial Information

Specific revenue data is not available publicly for Iridian, however the company said that revenues for fiscal 2002 will exceed revenues for the fiscal 2001 by over 100%.

The company is currently preparing for a C round of financing, which it expects to officially begin in September 2002. In 2000, Iridian Technologies completed a Series B funding raise. GE Equity, Perseus LLC, Joh Enschedé Security, Itochu, Walden Capital and Penny Lane investments are parties to the US$33.3 million raise.

4.10 Keyware Technologies SA/NV

Joseph Wybranlaan 40
Brussels B-1070
Belgium
Tel: +32 2 526 16 16
Fax: +32 2 524 38 67
Email:
info@keyware.com
Web:
www.keyware.com
General Manager:
Chris Buyse

4.10.1 History

Keyware Technologies was an early pioneer in the field of biometrics. Keyware was founded in 1996 and went public on 23 June 2000 on the pan-European technology exchange, NASDAQ Europe. More recently the company has changed its focus, due to difficulties it faced in the biometrics market.

4.10.2 Profile

Keyware had been a leading provider of biometric-enabled applications for user authentication for IT security, physical access, time and attendance, customer loyalty, electronic payment and ticketing. Over the years, it acquired multiple companies. Most of these acquisitions were vertical forward integrations aimed at speeding up the process of biometric deployment.

The company has recently redefined its strategy. Co-headquartered in Belgium and Woburn, Massachusetts, USA, the company employs 244 people and has offices in the USA, Belgium, Spain, Switzerland and France.

4.10.3 Technology

Keyware is best known for its layered biometric approach. As a form of biometric middleware, the company's layered biometric verification solution is device and vendor independent and can operate using face, voice and fingerprint recognition technologies. Keyware believes that a layered approach lessens a company's reliance on one single technology.

The company's main product is the Common Authentication Security (CAS) platform, which allows companies to centrally manage biometric and non-biometric (PIN, password, digital certificate) authentication methods.

4.10.4 Strategic Information

From being a biometric-centred company, Keyware has altered its strategy to become a provider of multi-application smart card solutions with a 'competence in biometrics'. It will focus on customer loyalty, electronic payment, event ticketing and identity application markets.

Keyware will be branded as a smart card solution provider. Its subsidiaries ABLE (internet security appliances) and TechTalk (Keyware's ID Division) will go to market with their own separate company brand name. Both ABLE and TechTalk are profitable.

Keyware provides its products through a combination of direct or indirect sales channels, depending on the geographic location of the customer. In Belgium and Luxembourg, Keyware mainly sells directly to customers. Through this direct interfacing with customers, Keyware is able to keep control over the quality of implementations and get feedback information about its customers' needs as well as the market evolution.

Outside this region, for both cost and efficiency reasons, Keyware tends to leverage the strengths of local partners and lines up with qualified and proven system integrators and value added resellers.

4.10.5 Financial Information

Combined with the sale of a number of non-core activities, Keyware increased its cash position to approximately US$4.9 million in the third quarter of 2002 – the proceeds of a private placement, in the form of a bond with warrants, amounted to US$1.9 million and divestments amounted to US$3 million. These funds should be enough to secure the cash needed to execute its updated business plan.

Keyware is confident that it will achieve positive cash flow towards the end of 2002. In 2001 operating losses at the company hit US$43.5 million leading to a 30% reduction in the workforce.

Keyware Technologies NV
Key Figures (US$ million) Year ended 31.12

	2001	2000	1999
Revenues	21.4	22.9	3.9
Net Loss	(5.3)	(10.1)	(43.6)

4.11 Nuance Communications Inc

1005 Hamilton Court
Menlo Park
CA 94025
USA
Tel: +1 650 847 0000
Fax: +1 650 847 7979
Email:
info@nuance.com
Web: www.nuance.com
President & CEO:
Ronald Croen

4.11.1 History

Nuance Communications Inc was established in 1994 as an independent company by Stanford Research Institute (SRI) International, one of the leading voice technology research entities throughout the 1980s and 1990s.

The company went public in April 2000 on the US Nasdaq. The supplier's *Verifier v 2.0* product was released in 2000, while version three was released in November 2001.

Today the company is headquartered in Menlo Park, California and has offices around the world.

4.11.2 Profile

Nuance's software incorporates speech recognition, text-to-speech synthesis and voice authentication technologies. The software powers a range of applications including automated directory assistance, voice-activated dialling, self-service banking and voice-driven customer relationship management.

The company also provides services such as system development, maintenance and training, which account for about half of sales. Customers include Nortel Networks (about 20% of sales), Sears and Charles Schwab.

4.11.3 Technology

The company's software servers run on industry-standard hardware and perform speech recognition, natural language understanding and voice authentication/ verification.

The company's main speaker verification product is Nuance Verifier 3.0. This product enables callers to be identified and authenticated simultaneously based on their voiceprint. The combination of speech recognition and voice authentication provides an additional level of security while eliminating the need for passwords and personal identification numbers (PINs). Nuance Verifier 3.0 provides support for international languages, allowing companies to deploy voice authentication worldwide.

The software is a proven technology, as demonstrated recently by a number of global deployments, including Banco Bradesco, the largest bank in Brazil, the

UK Government's Intensive Supervision and Surveillance Programme for fighting crime, and other major financial institutions for access to personal accounts and information.

The popularity of the company's verification engine is clear by the number of companies wanting to integrate it into their own solutions, such as Aspect Communications, InterVoice-Brite, Mitel Networks, Nortel Networks, Premier Geografix and Vocent.

4.11.4 Strategic Information

Nuance is probably the leading speaker verification company in terms of sales revenue, but other suppliers, such as VoiceVault are hot on their heels. The company generates its sales directly and through third-party value-added resellers, OEMs and system integrators. It is the indirect sales that the company believes will generate the most revenue in the future. Nuance's five largest resellers based on revenue raised in the first six months of 2000, were Edify (a subsidiary of S1 Corporation), Omron Corporation, Periphonics (a Nortel Networks company), Syntellect and Syscom.

As well as continually working on the accuracy of its verification technique, a possible technical development for the future could be the introduction of a transparent verification solution.

The company's main competitors include IBM, VoiceVault, Lucent Technologies, SpeechWorks International, Philips Electronics and Locus Dialogue.

4.11.5 Financial Information

Nuance investors include Alloy Ventures, Asset Management, Mayfield Fund, Morgenthaler Ventures and USVP. Corporate investors include Cisco Systems, Intel, Motorola, SAIC, Siebel Systems, SRI International, Sun Microsystems and Visa International.

In terms of revenue the company does not release segmented information concerning its biometric product range. However, the general trend for the company's overall revenue looks healthy.

Nuance Communications Inc

Key Figures (US$ 000) Year ended 31.12

	2001	2000	1999	1998
Revenue	39,225	51,818	19,567	11,755
Net Losses	(110,365)	(23,474)	(18,474)	(6,938)

NB These figures reflect all revenue, biometrics product sales only form a portion.

4.12 Persay Inc

Persay Inc
234 Crossways Park Drive
Woodbury
New York
NY 11797
USA
Tel: +1 516 6777291
Fax: +1 516 6777197
Email: info@persay.com
Web: www.persay.com
CEO: Almog Aley-Raz

Persay Ltd
22a Raul Wallenberg St
Tel Aviv 69719
Israel
Tel: +972 3 7678666
Fax: +972 3 7678661

4.12.1 History

Persay Inc was founded in 2000 as a small high-tech spin off from Comverse Infosys (now Verint Systems Inc). Persay's technology was developed in stages, as part of Comverse's ongoing research. Voice verification technology was originally developed for the security industry, but in 1996 developers realised it could also be used to provide security for civilian applications.

Being a spin off from a well-known and respected company has been a bonus for Persay. Its system, which is designed as a platform capable of handling hundreds of thousands of customers, are ideally suited for many companies in Comverse's client base.

4.12.2 Profile

Persay, a subsidiary of Comverse Technology, is one of the leading players in voice verification systems for remote services. It has a corporate office in New York with an R&D facility and a regional office in Tel Aviv, Israel.

Persay's system provides an answer to the growing need for efficient caller verification methods in financial call centres, telecom and e-commerce operations. Persay's solution employs unique verification technology, while capitalising on Comverse's experience in providing telephony and internet systems for large organisations.

In essence, Persay's new voice verification software is designed to obviate the need for people using online services to key in usernames, passwords and complicated codes. Instead, the user's voice is recorded by the system and, each time an online service is used, their identity can be verified in a matter of seconds.

4.12.3 Technology

The company's system is based on a combination of software and hardware, including a multi-processor server and communications cards made by Comverse, and software from Persay. The system operates via regular telephone lines or lines that use IP telephony. A bank or other utility will install the system on its organisational network and the system will monitor incoming phone calls.

Caller identification is achieved by comparing the voice of a caller to a digital voice signature stored in the system's databank. In order to safeguard the system from imposters using a recording of a voice, or any other attack method, the system re-identifies the caller several times during the conversation. The software identifies the physical characteristics of the speaker's voice, as these are stored in a database. In the course of the conversation, the system processes the vocal characteristics and authenticates the speaker's identity. The actual identification is based on a mathematical analysis of certain parameters in the voice that the human ear cannot discern. The company claims that this renders the system immune to imitators.

The system's level of accuracy is a trade secret, but sources at Persay claim it is higher than 90%.

Persay's product – *Orpheus Speaker Verification Platform* – enables call centres, telecom companies, e-commerce companies and any company that relies on remote services to verify the identity of their users through two optional modes.

The company can choose to match a voice password (like the customer's own name or telephone number) or to use 'free speech mode' – in which the caller is verified using the speech in a natural conversation.

4.12.4 Strategic Information

Even though Persay's business strategy is based on the sale of the system to clients, the company is currently laying the infrastructure for a service centre that will itself provide the voice verification service to potential customers. This centre may even serve as a future foundation for a host of additional services that would be supplied by other organisations. Persay's executives are hoping to make the new service available to customers in Europe, the US and Asia early in 2003.

The supplier's main customers have included Lloyds TSB in the UK and First Direct Bank. Persay's main competitors include Nuance, Intervoice Brite, Loquendo, Configate and VoiceVault.

4.12.5 Financial Information

Revenue information is not available publicly. However, in April 2000 Comverse obtained an investment of US$5 million from Shrem, Podim, Kelner in exchange for a 25% share in the new company. Comverse owns the remaining 75%.

There is talk of a second funding round open to strategic and financial investors for US$7 million, to close by the end of 2002.

4.13 Precise Biometrics AB

Sweden
Dag Hammarskjölds v 2
SE – 224 64 Lund
Sweden
Tel: + 46 46 311 100
Fax: + 46 46 311 101
Email: info@precisebiometrics.com
Web: www.precisebiometrics.com
CEO: Christer Bergman

4.13.1 History

Christer Fåhraeus and engineers connected to Lund's Technical College and the Ideon research village in Sweden established Precise Biometrics AB in 1997. Intensive development work led to the launch of the Precise 100 product suite in the summer of 1999. Today the company has approximately 60 employees and its headquarters are in Lund, Sweden. Elsewhere the company has offices in Stockholm and a US subsidiary in Washington DC.

4.13.2 Profile

Precise Biometrics is a publicly owned company and is listed on the O-list of the Stockholm Stock Exchange. The company supplies fingerprint authentication

USA
8300 Boone Boulevard,
Suite 500
Vienna, VA 22182
USA
Tel: + 1 703 848-9266
Fax: + 1 703 832-0577
Email: infous@
precisebiometrics.com

systems. The range of products includes systems for access control to computers and networks and for building into devices, such as mobile phones and portable computers.

The company's technology permits the use of very small sensors for identification, which it sees as a condition for integrating fingerprint sensors into mass-market devices. The company is also well-known for its Match-on-Card technology, which allows the storage and matching of fingerprint templates on a smart card – enabling the fingerprint template to never leave the confines of the secure smart card chip.

4.13.3 Technology

Precise Biometrics' technology is based on pattern recognition, reading the complete fingerprint, including the curve of the fingertip, three-dimensionally. The amount of information measured is three to ten times greater than information measured by earlier readers, the company claims. Its fingerprint readers are based on Precise BioMatch and Precise Match-on-Card technologies, which are BioAPI compliant.

The company's product range includes the following:

- *Precise 100 MC* – Fingerprint and smart card reader identifies individuals by their fingerprints.
- *Precise 100 SC* – Fingerprint templates are stored on a smart card to minimise the risk of template theft. The company's Match-on-Card technology lets users both store and match templates on the card.
- *Precise 100 A* – This is a very small fingerprint reader designed for the desktop. Fingerprint information is stored and matched on a hard drive or a server.
- *Precise 100 PC-card* – This is a solution for laptop users, with a pop out fingerprint sensor.
- *Precise BioKeyboard* – This is an ergonomic keyboard with a built-in fingerprint reader – the 100 A or 100 SC.
- *Precise BioAccess MC* – This is a plug and play fingerprint reader that offers security for access control. It uses templates stored in a smart card.
- *Precise BioAccess Mifare* – This is a plug and play fingerprint reader that uses templates stored in a Mifare contactless smart card and is designed for access control scenarios. At the heart of the solution is SecuriCore, an image processing and matching ASIC that can be integrated in card readers and electronic locks.
- *Precise BioMatch™ Pro Toolkit* – This is a development tool to integrate the Precise Match-on-Card technology into a software application that uses smart card and requires identification of a user.

4.13.4 Strategic Information

In 2001, Precise Biometrics restructured its organisation to better match market requirements. The intention was to increase the emphasis on customers and to react to the trend that sees physical access control and IT security solutions gradually converging. The company now has three different business areas: IT

Security, Physical Access and Embedded Solutions, while it has business units for the US Government, Nordic Key Accounts, IT Security Partners, Security Industry, Mobile Security and Smart Cards. The company markets, distributes and sells its products through extensive worldwide solution and channel partnerships.

The company has three patents, two of which were granted in 2001. In addition, four new patents were submitted in 2001. The company now has 27 patent applications referring to 13 different inventions. Notably, the company took out patent insurance in 2002 to reduce the company's economic risk should it be involved in a patent dispute. The company recently bought intellectual property from Veridicom in the USA.

Precise Biometrics has partnered with a number of leading application software providers, smart card developers, system integrators, value-added distributors and sensor manufacturers. These include ActivCard, BioNetrix, Entrust, ISL, NSuite Technologies, Orga, Protocom Development Systems, Saflink, SmartTrust, Spyrus and Utimaco Safeware. The company's main competitors include players such as Identix, Digital Persona and BioScrypt.

4.13.5 Financial Information

The company has seen fairly strong sales growth over the last few years. Sales during 2001 totaled SEK14.1 million (approximately US$1.49 million) compared with SEK5.2 million (US$550,000) in 2000. In the first half of 2002 turnover was SEK7.3 million (US$770,000). The company said that second quarter sales had been hit by quality shortcomings (and hence product recalls) on its Precise BioAccess device for physical access control.

A loss of SEK67.5 million (US$7.12 million) was recorded in 2001 compared with a loss of SEK43.4 million (US$4.58 million) in 2000. At the end of 2001, the company had SEK38.7 million (US$4.08 million) in liquid assets.

After the close of 2001, a targeted issue of new shares valued at SEK30.5 million (US$3.22 million) was made to Swedish institutional investors.

4.14 Recognition Systems Inc

1520 Dell Avenue
Campbell
CA 95008
USA
Tel: +1 408 341 4135
Fax: +1 408 341 4101
Email:
RSIsales@irco.com
Web:
www.recogsys.com
General Manager:
Martin Huddart

4.14.1 History

Recognition Systems Inc, a division of Ingersoll-Rand Company, specialises in biometric access control, time and attendance, and personal identification products. The company was founded in 1986 and has pioneered the commercialisation of biometrics using its patented hand geometry, which identifies people by the size and shape of their hand.

In 2000, the company expanded its corporate headquarters because of growing sales volumes. The new facilities feature additional manufacturing space, executive offices and a new training and conference centre for Recognition Systems' internal and external training seminars. The company now has approximately 40 employees.

4.14.2 Profile

The company is the world sales leader of biometric verification devices and serves an international clientele from its headquarters in Campbell, California. With over 55,000 hand geometry readers in use, Recognition Systems is the leader in hand recognition technology, which has developed to meet the growing needs of the commercial access control, time and attendance and identification markets.

Among the many installations worldwide, the company has enjoyed success in numerous high-profile applications. The 1996 Olympic Games in Atlanta, for example, used the HandReaders to protect access to the Olympic Village. More than 65,000 people were enrolled and over 1 million transactions were handled in 28 days. Recognition Systems HandReaders are also used increasingly in border crossing systems, banks, airports, prisons, universities, clubs and child care centres.

The widespread acceptance of hand geometry within access control and time and attendance markets has probably established Recognition Systems as the top supplier of biometric verification terminals.

4.14.3 Technology

Hand geometry scanners take over 90 measurements of the length, width, thickness, and surface area of the hand and four fingers in under one second. The technology uses a 32,000-pixel CCD digital camera to record the hand's three-dimensional shape from silhouetted images projected within the scanner. The scanner disregards surface details, such as fingerprints, lines, scars, and dirt, as well as fingernails, which may grow or be cut from day to day.

When a person uses the scanner, it compares the shape of the user's hand to a template recorded during an enrolment session. If the template and the hand match, the scanner produces an output – it may unlock a door, transmit data to a computer, verify identification, or log the person's arrival or departure time.

During enrolment, the image of a person's right hand is recorded three times and readings are taken for more than 90 physical characteristics. The average value for each measurement taken during the three scans is calculated. These average values create the 9-byte mathematical template for that person. Whenever the individual uses any HandReader, the image is compared to the template. The unit's internal memory can hold more than 32,000 different templates.

The company's product range includes:

- *Time and Attendance HandPunch Series* – A time and attendance terminal.
- *Access Control HandKey Series* – Provides door control and monitoring. Each HandKey II is a complete door controller providing door lock

operation, request for exit, and alarm monitoring. All information, including biometric data and decision-making capability resides locally.

- *HandNet Software* – Software that allows up to 120 HandKeys to be networked using a central PC.

4.14.4 Strategic Information

Recognition Systems is one of the biometric industry's most successful companies and this position is likely to remain for a while yet. The company attributes the majority of its success, profitability and growth to three specific markets: access and entry control; time and attendance; and personal identification (via visas, passports and other documents).

The rugged nature of the company's product makes it ideal in dirty, dusty environments, but the supplier is likely to introduce new variations of its hand geometry products, to be targeted at specific applications. As an example, the supplier integrated facial recognition into some of its readers earlier this year. The company has gained respect from its peers as a company that has many hundreds of successful implementations under its belt. As the physical access control market is still a dominant segment of the biometrics industry, the company is set to continue its dominant position.

Competition comes from one other company in the hand based recognition space – Biomet Partners – but more vigorous competition is from fingerprint and facial recognition companies.

Recognition Systems markets its products domestically and internationally through system integrators, distributors, dealers and selected business partners and OEMs.

4.14.5 Financial Information

Although the majority of the company sales come from North American markets, the company has a presence in all the other main geographic regions.

As a division of Ingersoll-Rand, specific biometric financial information is not available publicly. However, the company was certainly one of the largest in the industry today (in the top three). The company expects steady, but not exceptional, growth in the future, which means the supplier will probably be overtaken by numerous other companies in the next two to three years. This will occur as IT security markets pick up – an area which cannot be tapped by bulky hand reader technology.

4.15 SecuGen Corporation

348 Montague
Expressway
Milpitas
CA 95035
USA
Tel: +1 408 942 3400
Fax: +1 408 262 0264
Email:
sales@secugen.com
Web: www.secugen.com
President & CEO:
Andy Szego
CEO: Robert Kyle

4.15.1 History

Founded in 1998, SecuGen Corporation (short for Secure Generation) is one of the main players in the optical fingerprint recognition market. Headquartered in the USA, the company has a number of locations worldwide including offices in Ontario, Canada (Secugen Canada), Tokyo, Japan (Secugen Japan Ltd), and in Hong Kong (Secugen Asia Ltd).

4.15.2 Profile

SecuGen is a privately held company. It designs, manufactures and markets its optical fingerprint sensor technology to create biometric solutions for a range of applications and industries including airports, enterprise networks, healthcare, e-business, banking, time and attendance, physical access and gaming.

SecuGen's products are sold as a complete solution, an OEM component, or integrated into ready-to-use PC peripherals such as optical mice, keyboards and a standalone desktop device. Partnerships with key network security software suppliers assure compatibility across many different platforms and applications.

4.15.3 Technology

SecuGen's core technology is an optical sensor module combined with a set of extraction and matching algorithms developed for use with the company's SEIR (Surface Enhanced Irregular Reflection) optical method. The module is embedded in SecuGen's line of fingerprint PC peripheral devices and stand-alone devices produced by OEMs for applications such as access control and time and attendance. Software is available for desktop computer users, networks and the internet.

The company claims that its SEIR technology helps to capture images that are undistorted and of high resolution. SecuGen devices also reportedly protect against latent and "faked" fingerprints. The problem of faked fingerprint hit the industry hard earlier this year and Secugen responded saying it has now made physical changes to its sensor and added a new proprietary algorithm, which together can defeat all known attack methods.

The company claims that its patent-pending scratch-proof platen of the optical module is as hard as quartz.

SecuGen's product range includes the following:

Devices

- *SecuGen Hamster II*
- *SecuGen Keyboard – with optional smart card reader/writer*
- *SecuGen Mouse*
- *SecuGen OptiMouse*

Software

- *SecuDesktop2000* – SecuGen's fingerprint biometric security solution for desktop/ laptop workstations. The fingerprint-activated 128-bit encryption algorithm provided with SecuDesktop 2000, secures specified data files. For networked computers, the domain set-up feature enables registered users to login to any Windows domain to which they belong.
- *SecuIBAS* – A product that allows secure internet transactions.

Fingerprint sensors

- *FDA01* – Standalone sensor not requiring PC connection
- *FCA01* – Standalone sensor not requiring PC connection (flex version)
- *FDP02* – PC-connectible for parallel port
- *FDU01* – PC-connectible for USB port

Other

- The company has different developer kits as well as OEM products for physical access control and time/attendance etc.

4.15.4 Strategic Information

As a provider of peripherals and sensors it is vitally important to make its products as compatible and easy to use as possible as well as ensuring widespread available. The company has made necessary moves to ensure interoperability with leading players such as Novell and Microsoft.

An important part of SecuGen's strategy is to establish widespread channels for the distribution of its products. In order to achieve this the company sells its products through an extensive network of value added resellers, independent software vendors (ISV) and OEMs, which it is continually attempting to develop.

SecuGen's ISV partners provide value added resellers and systems integrators with enterprise solutions that support most available LDAP directories including Microsoft's Active Directory and Novell's NDS.

In particular the company wants partners that target the financial services, healthcare, physical access control/time and attendance, and government market segments. In order to tempt these types of partners SecuGen offers a fairly good margin on peripherals sold.

Another priority in the company's strategic plan is the support of industry standards, such as BioAPI and CBEFF. It demonstrates this support through membership of the IBIA and the BioAPI Consortium.

4.15.5 Financial Information

As a private company SecuGen does not release any financial data publicly.

4.16 STMicroelectronics

1000 East Bell Road
Phoenix
AZ 85022
USA
Tel: +1 602 485 6100
Fax: +1 602 485 6102
Web: www.st.com
President & CEO:
Pasquale Pistorio

4.16.1 History

The ST group was formed in June 1987 as a result of the merger between SGS Microelettronica of Italy and Thomson Semiconducteurs of France. In May 1998, the company changed its name from SGS-THOMSON Microelectronics to STMicroelectronics. STMicroelectronics is quoted on the New York Stock Exchange, the Euronext Paris, and since June 1998, on Borsa Italiana.

STMicroelectronics' main biometric division is clearly only a very small part of the overall group. Since it began operations the division acquired software development group Skytale from US supplier Veridicom in 2001. The deal included the mechanical design of peripherals and the source code of the supplier's Protector Suite algorithms, which it now sells along with its sensor range.

4.16.2 Profile

A global independent semiconductor company, STMicroelectronics is a leader in developing and delivering semiconductor solutions across the spectrum of microelectronics applications, including biometrics. The group totals about 40,000 employees, 12 advanced research and development units, 32 design and application centres, 17 main manufacturing sites and 74 sales offices in 27 countries.

STMicroelectronics' main biometric division is based in the USA. STMicroelectronics claims to have shipped over 100,000 units in its first year of production and an additional 50,000 since then. The company claims that this level of deployment makes it the most industrialised of the silicon sensor manufacturers. However, it is likely that other major sensor manufacturers are not far from this level, as the interest in silicon based sensors increased across the board in the last two years.

4.16.3 Technology

The biometric division's main technology centres round the TouchChip range of area sensors. TouchChip is an active capacitive pixel-sensing technology. Each sensor cell (pixel) contains an active capacitive feedback circuit whose effective feedback capacitance is modulated by the presence of live skin close to the surface of the sensor. This active sensing approach provides a higher immunity to parasitic effects. This, the company claims, ensures a higher signal-to-noise ratio and greater capacity to capture a wider range of fingerprints than passive capacitive sensing.

The sensors produced by STMicroelectronics are:

- *TCS1A* – This has an active area of 12.8 × 18 mm (one of the largest on the market) and an image resolution of 508 dpi. It is ideally suited for high security applications.

- *TCS2A* – This has an active area of 10.4 × 14.4 mm and an image resolution of 508 dpi. It is designed for portable devices.

Its reader family comprises:

- *TCRU1A* – TouchChip Fingerprint USB reader
- *TCRS1A* – Combined Fingerprint and Smartcard USB reader
- *TCRA1A* – Pocket-Sized Fingerprint and SIM Smartcard USB reader

The company's biometric security software is called *TouchChip ProtectorSuite*. The company also has embedded systems and SDKs.

4.16.4 Strategic Information

Silicon sensor technology has taken off in the last two years. Many of the challenges, such as physical robustness, resistance to electrostatic discharge and price have been addressed, and the sensors are now in a position to compete strongly for market share against optical based fingerprint sensors. In this growing market STMicroelectronics is probably the market leader closely followed by the likes of Infineon, Atmel and AuthenTec.

Similar to these other companies, STMicroelectronics is likely to be suffering from the global slowdown in IT security markets, which have been affected by economic tribulations – particularly in the USA, where STMicroelectronics has its base.

From a technical standpoint, the company sees the value in developing a swipe-based sensor to fit into devices, such as mobile phones and PDAs. The company has a swipe sensor in development and an integrated ASIC/sensor design, which is expected to be released in 2002.

Another of the company's overall strategies is to combine navigation with authentication – so instead of using buttons on a mobile phone to navigate the screen, a swipe sensor would be used.

4.16.5 Financial Information

In 2001, STMicroelectronics's overall net revenues were US$6.36 billion and net earnings were US$257.1 million, putting it in the top three semiconductor companies in the world.

There is no specific breakdown of its TouchChip division, but the number of sensors sold in its first year does give an indication of revenue generation (100,000 sensors at an estimated US$50 would give revenues of US$5 million).

4.17 Viisage Technology Inc

30 Porter Rd
Littleton
MA 01460
USA
Tel: +1 978 952 2200
Fax: +1 978 952 2225
Email:
info@viisage.com
Web: www.viisage.com
CEO & President:
Bernard Bailey

4.17.1 Company History

Viisage Technology Inc began operations in 1993 as a division of Lau Technologies. In November 1996, Viisage completed its initial public offering, and Lau retained a 64% holding. In June 1998, Viisage created a separate biometrics division, focusing on product, market and channel development in three main areas: facility access control; PC network and internet access security; and real-time large database identification and verification of individuals.

Viisage has recently completed the acquisition of Biometrica Technology, a leader in facial recognition systems for the casino industry, having over 150 casino customers. The company also bought out Lau Security Systems in January 2002. In return, Viisage will pay Lau a royalty of 3.1% of facial recognition revenues over a 12.5 year period, up to a maximum of US$27.5 million.

More recently Viisage acquired a substantial part of the Miros division of eTrue.com – which has shut its doors for business. The deal gave Viisage access to a base of customers with more than 400 installations in the European market as well as to Miros's TrueFace software.

4.17.2 Company Profile

Viisage Technology specialises in digital identification solutions. Originally developed at the Massachusetts Institute of Technology, Viisage's face-recognition technology, systems integration and software design capabilities provide identity protection solutions for security and e-commerce applications. Early adopters are using the company's face recognition software for customer ID verification at ATMs, internet verification for e-commerce and home workers, large database fraud reduction, casino surveillance, airport and other security and law enforcement applications.

Every year Viisage systems produce more than 25 million digital identification documents in the USA through 1200 locations. Applications include driver licenses, social security cards and law enforcement credentials. Viisage has contracts in 16 US states, although just four use facial recognition capabilities at present.

Internationally the company has designed and implemented voter registration and national ID systems in several countries.

4.17.3 Technology

Viisage's patented face-recognition technology uses a sophisticated algorithm based on principal component analysis. The company's software converts the characteristics of the face into a unique set of numbers known as an "eigenface". The technology was originally developed at MIT, but additional research (over 100 person years, the company claims) has taken the technology beyond its original capabilities.

Viisage's face recognition operates in the following scenarios: e-commerce *(FacePASS)*; point-of-sale or ATMs *(FacePIN)*; access control *(FacePASS)*; surveillance *(FaceFINDER)*; image database analysis *(FaceEXPLORER)*; software development kit *(FaceTOOLS)*; and image quality *(FaceADVISOR)*.

4.17.4 Strategic Information

Since the terrorist attacks of September 11th, the market landscape has changed dramatically – creating new demands for face-recognition technology to aid in national security. Government contracts, such as driver licenses, and programs for armed services and other agencies, constitute a large portion of Viisage's business. The competition for such contracts is intense, however, and the process is often lengthy. Viisage believes it has positioned itself to respond to these large emerging market opportunities, with one of its three business units specifically focused on Homeland Defense. The company's other two divisions are Commercial Solutions (for corporate customers, including the gaming industry) and 3SI (for smart and secure identification solutions).

While Homeland Security is complex and multifaceted, the most fundamental element to security is identification – and the company has already had success in securing a number of deals, notably in the driver license and airport sectors. Viisage's technology has been deployed (mostly as trials) in Fresno Yosemite International Airport in California, Manchester Airport in New Hampshire (system now purchased), St Petersburg-Clearwater International Airport in Florida (purchased) and Boston's Logan International Airport. As part of the system provided to the airports, Viisage introduced SecureSITE, a free standing, automated image capture system that is integrated into Viisage's biometric networked architecture (BNA) overt airport security screening.

The company fully supports a Bill – The Driver's License Modernisation Act 2002 – which has a provision to include biometric identifiers within a new standardised document. However this support is unsurprising considering Viisage already provides around one third of the USA's driver licenses. If its facial recognition technology was to be included in the driver licenses (as is already the case in four of its state customers), then this would provide a clear financial boost to the company's revenue performance.

Elsewhere the company has recently been awarded orders by law enforcement departments, casinos and the Department of Defense for battlefield uses as well as for military security. Internationally Viisage's technology was used for a more than US$1 million project with the Electoral Commission of Uganda to deter voter fraud. The system reportedly identified more than 30 specific voters who were registered more than once.

On the merger and acquisition front Viisage intends to pursue strategic opportunities to accelerate its growth in every market segment – this has been seen recently in the acquisition of Biometrica and part of eTrue.com. The company plans to complement its acquisition strategy with organic growth by continuing to invest in new technologies, such as specific biometric cameras, that are expected to drive down costs for surveillance and access control applications.

The company has formed important strategic partnerships with Unisys Corporation and Curtiss-Wright, while it formed a distribution channel for face recognition in the Middle East through a partnership with Silver Network Electronic Co of Kuwait. In the facial recognition arena the company's biggest competitor is Identix.

4.17.5 Financial Information

Viisage's public figures show that its facial recognition activities are growing fast. In 2000 biometrics accounted for approximately 10% of its overall sales, a figure that has risen to 20% in the first six months of 2002. In dollars this was approximately US$6.17 million in the first half of 2002 compared with US$3.68 million in the whole of 2001, US$2.47 million in 2000 and just US$120,000 in 1999.

4.18 Voicevault (formerly Buytel)

Clonmel House
17 Harcourt Street
Dublin 2
Ireland
Tel: + 353 1 603 9500
Fax: + 353 1 478 4478
Email:
info@voicevault.com
Web:
www.voicevault.com
CEO: Patrick Keaney

4.18.1 History

Voicevault, formerly known as Buytel, was founded in 1996 and is a privately owned and funded company.

The original service that was developed was a purchasing-by-phone service called Teledirect. This is now just one of a number of applications and voice verification is the method of accessing any of these applications.

Voicevault owns all of its intellectual property. In May 2001, Voicevault acquired the 'Speakerkey' engine, an integral part of its voice verification technology, from the US based ITT Industries. Up until then, Voicevault had licensed the Speakerkey engine from ITT on an exclusive worldwide license agreement. Since then, Voicevault has furthered the ITT technology, significantly improving the speed and increasing the rate of accuracy and reliability.

In April 2002, Voicevault was accredited with BS 7799, the first company to receive this standard in Ireland.

4.18.2 Profile

Voicevault is headquartered in Dublin, Ireland, has offices in the UK, Brazil, South Africa, Sweden and the US, and is currently in the process of setting up an office in Japan. Voicevault, which is one of the leading companies in the application of voice verification, verifies people by their voice over the phone or over the internet in less than a second regardless of where they are located.

4.18.3 Technology

Voicevault's voice verification solution offers a replacement and/or an enhancement for PINs, passwords and other 'Mother's Maiden name?' routines. The key benefits of Voicevault include cost and fraud reduction, increased security, improved efficiency and enhanced overall customer experience. Voicevault claims to be the best in terms of speed, accuracy and value; performing a verification to anywhere in the world in less than a second and offering an accuracy level in the high ninety percentage range.

Voicevault has developed a range of commercial applications including:

- *Password Reset* – For the secure, automated resetting of passwords and PINS without the need for helpdesk intervention.
- *Secure Access Management* – Allows employees to gain access to confidential company information over the phone or the internet in a secure and user-friendly manner.
- *SSM* – Designed to provide voice-based access to a secure session key and to include voice verification to authenticate access to that key ('voice PKI').
- *Call Centre Gateway* – Is a more efficient and user-friendly method of accessing call centres, which also increases the productivity of call centre agents.
- *Buywayz Voice Purchasing Portal* (www.buywayz.com) – Provides end-to-end m-commerce solutions for use in sales and/or purchasing processes.

In terms of the architecture, Voicevault services are available in three different configurations: *Voicevaultphone, Voicevaultnet, Voicevaultweb*, depending on the application and location of the server.

4.18.4 Strategic Information

Voicevault has spent the last few years developing its product and applications, securing its IP and building up brand awareness. It is now concentrating on building up its existing references. It has two dedicated R&D teams, one based in Dublin and the other in San Diego, both of which concentrate heavily on the development of new applications in accordance with its market analysis team. The company is developing a number of new applications in conjunction with systems integrators that are now close to completion.

Voicevault originally operated purely on the basis of an ASP model. However the company has now adopted a more flexible approach and its voice verification technology is available as an ASP model, utility or license based model, depending on the requirements of the client.

In addition to selling directly through its sales force, Voicevault also has a number of reseller agreements in place, Accenture being one of the most notable. Voicevault also has partnership agreements in place in each of its key markets, namely Europe, the Americas and Africa. It is currently in the advanced stages of negotiation to sign three partnership deals with major global systems integrators.

Voicevault has an extensive range of global clients in the financial services, healthcare, public services, communications and high tech sectors. Among its customers are AIB, a leading European financial services organisation; Associated Mediabase, the technology arm of Associated Newspapers Limited; Telia, a global telecommunications services company; Progressive, one of the largest insurance writers in the US; Trintech, a player in the secure payments sector and Group 4, a player in the community corrections area.

Voicevault has very recently signed a major deal with one of the largest high street retail banks in the UK.

4.18.5 Financial Information

In May 2001, Voicevault completed its second round of funding, raising €15 million led by Accenture Technology Ventures with West LB Panmure as co-lead. As Voicevault is a privately owned company, no other financial information can be disclosed.

5 Address Book

5.1 Alphabetical Company Listing

3M-AiT
1545 Carling Avenue, Suite 700, Ottawa, Ontario, Canada K1Z 8P9
Tel: +1 613 722 2070
Fax: +1 613 722 2063
Email: marketing@ait.ca
Web: www.ait.ca
President & CEO: Bernie Ashe

A

A4Vision
26 rue Maunoir, 1207 Geneva, Switzerland
Tel: +41 22 849 1050
Fax: +41 22 849 1060
Email: kelly.richdale@a4vision.com
Web: www.a4vision.com
CEO: Kelly Richdale

Access Controls Inc
1064 NW 53rd Street, Fort Lauderdale, FL 33309, USA
Tel: +1 954 772 2202
Fax: +1 954 771 5706
Email: info@acizone.com
Web: www.acizone.com
President & CEO: Jim Gordon

Accu-Tech Systems Ltd
C1, Caerphilly Business Park, Van Road, Caerphilly, Mid Glamorgan CF83 3ED, UK
Tel: +44 2920 885599
Fax: +44 2920 868666
Email: enquire@accu-tech.co.uk
Web: www.accu-tech.co.uk
Sales Director: David Hembery

Accu-Time Systems Inc
420 Somers Road (Rt. 83), Ellington, CT 06029, USA
Tel: +1 860 870 5000
Fax: +1 860 872 1511
Email: sales@accu-time.com
Web: www.accu-time.com
President: Peter DiMaria

AcSys Biometrics Corp
399 Pearl Street, Burlington, Ontario, Canada L7R 2M8
Tel: +1 905 634 4111
Fax: +1 905 634 1101
Email: info@nxsgrp.com
Web: www.acsysbiometrics.com
President: Jerry Janik

ActivCard Inc
6623 Dumbarton Circle, Fremont, CA 94555, USA
Tel: +1 510 574 0100
Fax: +1 510 574 0101
Web: www.activcard.com
CEO: Steven Humphreys

Aditech Limited
International House, Aylesbury, Buckinghamshire HP19 8BP, UK
Tel: +44 1296 398085
Fax: +44 1296 337755
Email: sales@aditech.co.uk
Web: www.aditech.co.uk
Contact: Paul Stanborough

Advanced Biometrics Inc
7203 W Deschutes Ave, Kennewick, WA 99336, USA
Tel: +1 509 736 7000
Fax: +1 509 736 7001
Email: sales@livegrip.com
Web: www.livegrip.com
CEO: James C Katzaroff

Advanced Software Technologies (AST)
Calle Caleruega 102, Madrid 28033, Spain
Tel: +34 9 1 766 46 00
Fax: +34 9 1 383 00 59
Email: ast@ast-afis.com
Web: www.ast-afis.com

Affinitex
700-1545 Carling Avenue, Ottawa, Ontario, Canada K1Z 8P9
Tel: +1 613 722 6353
Fax: +1 613 722 8378
Email: info@affinitex.com
Web: www.affinitex.com

AGMA Co
Domaine Technologique de Saclay, 4 rue René Razel, Orsay Cedex 91892, France
Tel: +33 1 69 41 32 38
Fax: +33 1 69 41 35 23
Email: agma-morpho@wanadoo.fr
Web: http://perso.wanadoo.fr/agma-morpho

AIMS Technology Inc
PO Box 1093, Little River, SC 29566, USA
Tel: +1 843 399 2202
Fax: +1 843 399 0312
Email: aims@nail-id.com
Web: www.nail-id.com
Contact: Austin Gormley

ALMEX Ltd
3853 Trelawny Circle, Mississauga, Ontario, Canada L5N 6S4
Tel: +1 905 785 0855
Fax: +1 905 785 2811
Email: sales@almexltd.com
Web: www.almexltd.com/access1.htm
Contact: Vesna Podrug

Alternative Computer Technology
7908 Cincinnati-Dayton Road, Suite W, West Chester, OH 45069, USA
Tel: +1 513 755 1957
Fax: +1 513 755 1958
Email: sales@altcomp.com
Web: www.altcomp.com

AND Corporation
55 University Ave, Toronto, Ontario, Canada M5J 2H7
Tel: +1 416 920 8260
Fax: +1 416 920 8409
Email: sutherland@andcorporation.com
Web: www.andcorporation.com
President: John Sutherland

Anovea Authentication Technology Inc
1797 Spruce Street, Berkeley, CA 94709, USA
Email: info@anovea.com
Web: www.anovea.com

App Informatik Davos
PO Box 7270, Rosenhuegel 6, CH 7270 Davos, Switzerland
Tel: +41 81 410 09 90
Email: info@app-davos.ch
Web: www.app-davos.ch

Applied Biometrics Products Inc
600 Blair Park Road, Suite 155, Williston, VT 05495, USA
Tel: +1 802 872 2760
Fax: +1 802 872 3228
Email: appliedbiometrics@usa.net
Web: www.appliedbiometrics.net
Chairman & CEO: Dr John F Catalano

Arithmetic Guarding Equipments (AGE)
8F, 66 Fu Hsiang Rd, Chung Ho City, Taipei Hsien, Taiwan
Tel: +886 2 82421868
Fax: +886 2 82421865
Email: sales@age2001.com
Web: www.age2001.com

Association for Biometrics
AfB Secretariat, 3 Ravens Croft, East Hunsbury, Northampton, Northamptonshire NN4 0RL, UK
Tel: +44 1604 660 125
Fax: +44 1604 709 855
Email: info@afb.org.uk
Web: www.afb.org.uk
Chairman: Gerry Gardner

Astro Datensysteme AG
Am Bleichbach 28, D-85452 Moosinning, Germany
Tel: +49 8123 926 0
Fax: +49 8123 926 290
Email: vertrieb@astro.de
Web: www.astro.de
Contact: Stephan Obermaier

ASTRONTECH Sp
Atrium Plaza, Al. Jana Pawla II 29, Warszawa 00-867, Poland
Tel: +48 22 528 91 09
Fax: +48 22 528 93 09
Email: info@astrontech.pl
Web: www.astrontech.pl

Atmel Corporation
2325 Orchard Parkway, San Jose, CA 95131, USA
Tel: +1 408 441 0311
Email: misc@atmel.com
Web: www.atmel.com
President & CEO: George Perlegos

Audata Ltd
Willesborough Industrial Park, Kennington Road, Ashford, Kent TN24 0TD, UK
Tel: +44 1233 660 243
Fax: +44 1233 640 690
Email: info@audata.co.uk
Web: www.audata.co.uk

Audiopoint
3998 Fair Ridge Drive, Fairfax, VA 22033, USA
Tel: +1 703 279 4400
Web: www.myaudiopoint.com
President & CEO: Brian L Lichorowic

Aurora Biometrics Inc
5 Choke Cherry Road, Suite 360, Rockville, MD 20850, USA
Tel: +1 301 590 0711
Fax: +1 301 926 6940
Email: pchesson@AuroraBiometrics.com
Web: www.aurorabiometrics.com
Chairman & CEO: Dr Helena S Wisniewski

Aurora Computer Services Ltd
PO Box 5579, Northampton NN6 9GG, UK
Tel: +44 870 606 0412
Fax: +44 1604 780 801
Email: info@auroracs.co.uk
Web: www.auroracs.co.uk

AuthenTec Inc
PO Box 2719, 709 S Harbor City Blvd, Suite 400, Melbourne, FL 32902-2719, USA
Tel: +1 321 308 1300
Fax: +1 321 308 1430
Email: mkt@authentec.com
Web: www.authentec.com
President & CEO: Scott Moody

Automa Pty Limited
PO Box 320, Ryde, NSW 1680, Australia
Tel: +61 2 9808 6939
Fax: +61 2 9807 1577
Email: sales@automa.com.au
Web: www.automa.com.au
Contact: Shaun Werkhoven

Autostar Technology
Blk 5012 Ang Mo Kio Ave 5, Unit 04-10, TECHplaceII, Singapore 569876
Tel: +65 6482 9595
Fax: +65 6482 1833
Email: sales@autostar.com.sg
Web: www.autostar.com.sg
Managing Director: KG Koh

Avalanche Technologies (Pty) Ltd
PO Box 13794, Lois Ave 509, Hatfield 0028, South Africa
Tel: +24 12 3434024
Fax: +24 12 3444892
Email: info@avalanche.za.com
Web: www.avalanche.za.com
Contact: Eugene Smit

Aware Inc
40 Middlesex Turnpike, Bedford, MA 01730, USA
Tel: +1 781 276 4000
Fax: +1 781 276 4001
Email: info@aware.com
Web: www.aware.com

Axis Software
1202, Bhosale terrace, Off Apte Road, Shivajinagar, Pune 411004, India
Tel: +91 20 5530297/8
Fax: +91 20 5537591
Email: info@axistech.com
Web: www.axistech.com
CEO & President: Abhay Khinvasara

Axxess Identification Ltd
8 Shrivenham Hundred Business Park, Watchfield, Swindon, Wiltshire SN6 8TZ, UK
Tel: +44 1793 784002
Fax: +44 1793 784005
Email: info@axxessid.com
Web: www.axxessid.com
Contact: Sandra Glover

B

Bannerbridge
22-24 Hornsby Square, Southfields Industrial Park, Basildon, Essex SS15 6SD, UK
Tel: +44 870 870 1220
Fax: +44 870 870 1202
Email: security@bannerbridge.co.uk
Web: www.bannerbridge.co.uk
Contact: Neil Kinsman

Banque-Tec International Pty Ltd
Unit 5, 12-18 Victoria St, East Lidcombe, NSW 2141, Australia
Tel: +61 2 9749 4999
Fax: +61 2 9749 5100
Email: info@banquetec.com
Web: www.banquetec.com
CEO: Chris Burke

Bell ID
PO Box 29141, Rotterdam 3001 GC, The Netherlands
Tel: +31 10 885 10 10
Fax: +31 10 885 10 11
Email: info@bellid.com
Web: www.bellid.com
Contact: Igno Peters

Bell Security
Roding House, 970 Romford Road, London E12 5LP, UK
Tel: +44 20 8553 5932
Fax: +44 20 8478 8052
Email: info@bellsec.com
Web: www.bellsec.com
Managing Director: David Orme

Bergdata Biometrics GmbH
Koblenzer Strasse 9, 53173 Bonn, Germany
Tel: +49 228 3680 110
Fax: +49 228 3680 118
Email: info@bergdata.com
Web: www.bergdata.com
Contact: Gerik von Graevenitz

Bioacesso
Av Miguel Bombarda, n°61 4°Esq, Lisbon 1050-161, Portugal
Tel: +351 21 312 96 20
Fax: +351 21 312 96 25
Email: info@bioacesso.com
Web: www.bioacesso.com
Contact: Miguel Damas de Matos

Biocentric Solutions Inc
8417 Excelsior Drive, Madison, WI 53717-1901, USA
Tel: +1 608 821 0821
Fax: +1 608 821 0822
Email: cschaub@biocentricsolutions.com
Web: www.biocentricsolutions.com
President & CEO: Martin Janiak

BioconX Inc
7400 Metro Boulevard, Suite 425, Minneapolis, MN 55439, USA
Tel: +1 952 835 5321
Fax: +1 952 835 7410
Email: info@bioconx.com
Web: www.bioconx.com
President & CEO: Vincent Hren

BioEnable Technologies
74A 1/1, Prakash Complex, Kawade Road, Ghorpadi, Pune, MH 411001, India
Tel: +91 20 663 3419
Email: info@bioenabletech.com
Web: www.bioenabletech.com

BioIdentix Inc
3802 Handberry Lane, Pearland, Houston, TX 77562, USA
Tel: +1 281 412 6670
Web: www.bioidentix.net
Contact: James Gries

BIO-key International
1285 Corporate Center Drive, Suite 175, Eagan, MN 55121, USA
Tel: +1 651 687 0414
Fax: +1 651 687 0515
Email: sales@bio-key.com
Web: www.bio-key.com
CEO: Jeffry Brown

BioLink Technologies International Inc
3949 Commerce Parkway, Miramar, FL 33025, USA
Tel: +1 954 450 7061
Fax: +1 954 450 7062
Email: info@biolinkusa.com
Web: www.biolinkusa.com
CEO: Gene Chayevsky

BioLynx Inc
7550 IH10 West, Floor 14, San Antonio, TX 78229, USA
Tel: +1 210 256 8300
Fax: +1 210 682 2137
Email: sales@biolynx.com
Web: www.biolynx.com
Chairman & CEO: Alan W Saltzman

BioMet Partners Inc
Pestalozzistrasse 12, CH-3280 Morat, Switzerland
Tel: +41 26 670 13 03
Fax: +41 26 670 13 02
Email: mafsaswiss@compuserve.com
Web: www.biomet.ch

Biometri Systemer AS
Trondheimsveien 184, Oslo NO-0570, Norway
Tel: +47 95 25 78 45
Fax: +47 22 60 44 27
Email: mail@biometri.no
Web: www.biometrics.no
Contact: Thomas Braekke

Biometric Access Corp
2555 Interstate Highway 35, Suite 200, Round Rock, TX 78664, USA
Tel: +1 512 246 3760
Fax: +1 512 246 3768
Email: sales@biometricaccess.com
Web: www.biometricaccess.com
President & CEO: Ronald R Smith

Biometric Associates Inc
9475 Deereco Road, Suite 410, Timonium, MD 21903, USA
Tel: +1 410 252 7210
Fax: +1 410 252 7214
Email: sales@biometricassociates.com
Web: www.biometricassociates.com
President & CEO: Doug Kozlay

Biometric Consortium
Email: info@biometrics.org
Web: www.biometrics.org

Biometric Digest
PO Box 510047, St. Louis, MO 63151-0047, USA
Tel: +1 314 892 8632
Fax: +1 314 487 5198
Email: editor@biodigest.com
Web: www.biodigest.com
Editor: William Rogers

Biometric Partners Europe
Tel: +49 7325 5838
Fax: +49 7325 5802
Email: r_hauke@biometricpartners.com
Web: www.biometricpartners.com
Contact: Rudolf Hauke

Biometric Technology Today
Elsevier Advanced Technology, The Boulevard, Langford Lane, Kidlington, Oxford OX5 1AS, UK
Tel: +44 1865 843000
Fax: +44 1865 843971
Email: m.lockie@btopenworld.com
Web: www.biometrics-today.com
Editor: Mark Lockie

Biometrica Systems Inc
PO Box 430, Mont Vernon, NH 03057, USA
Tel: +1 603 673 6218
Fax: +1 603 673 6271
Email: info@biometrica.com
Web: www.biometrica.com
Contact: Bob Schmitt

Biometrics Direct
PO Box 403, Artemis Solutions Group LLC, Freeland, WA 98249, USA
Tel: +1 360 331 1071
Fax: +1 360 331 1072
Email: james_childers@biometricsdirect.com
Web: www.biometricsdirect.com
CEO: James Childers

Biometrics Imagineering Inc
252 Ponce de Leon Ave, Citibank Tower, 16th floor, San Juan, PR 00976-2001, Puerto Rico
Tel: +1 787 753 5323
Fax: +1 787 753 5340
Email: info@biocheck.com
Web: www.biocheck.com
Chairman, President & CEO: Héctor T Hoyos-Aliff

Biometrics.co.za
PO Box 51455, Waterfront, Capetown 8002, South Africa
Tel: +27 21 439 8086
Fax: +27 21 434 2916
Email: info@biometrics.co.za
Web: www.biometrics.co.za
CEO: Dr Harold Kimmel

Biometrika
via Bertini 75, Forlì 47100, Italy
Tel: +39 0543 798604
Fax: +39 0543 792064
Email: info@biometrika.it
Web: www.biometrika.it
Contact: Giorgio Alboni

Biometrix International
Hauffgasse 19, Vienna A-1110, Austria
Tel: +431 748 1756
Fax: +431 748 1756
Email: info@biometrix.at
Web: www.biometrix.at
General Manager: Harald Griesser

BioNetrix System Corporation
1953 Gallows Road, Suite 500, Vienna, VA 22182, USA
Tel: +1 703 734 9200
Fax: +1 703 734 6553
Email: info@bionetrix.com
Web: www.bionetrix.com
CEO & Chairman: Steve Walker

BioPay
580 Herndon Parkway, Suite 100, Herndon, VA 20170, USA
Tel: +1 866 324 6729
Fax: +1 703 435 3120
Email: webmaster@biopay.com
Web: www.biopay.com
President: Tim Robinson

Bioscrypt Inc
5450 Explorer Drive, Suite 500, Mississauga, Ontario, Canada L4W 5M1
Tel: +1 905 624 7700
Fax: +1 905 624 7742
Email: jean-pierre.lee@bioscrypt.com
Web: www.bioscrypt.com
President & CEO: Pierre Donaldson

Bio-Sec Corporation
Dufourstrasse 43, CH 8034 Zurich, Switzerland
Tel: +41 1 252 58 68
Fax: +41 1 252 58 68
Email: info@bio-sec.com
Web: www.bio-sec.com

BioThentica Corporation
11803 Grant Road, Suite 112B, Cypress, TX 77429, USA
Tel: +1 866 744 7226
Fax: +1 888 694 1995
Email: info@biothentica.com
Web: www.biothentica.com
Co-Founder & Chairman: David L. Tautfest

BioTrust Project
Chamissostrasse 11, D-99096 Erfurt, Germany
Tel: +49 6126 925 85
Fax: +49 361 34 53 957
Email: info@biotrust.de
Web: http://biotrust.de
Project Manager: Hans-Henning Arendt

BrainTrust Technology GmbH
Landsberger Allee 366, 12681 Berlin, Germany
Tel: +49 30 54 97 93 0
Fax: +49 30 54 37 57 4
Email: info@goldfinger-secure.com
Web: www.goldfinger-secure.com

BTG International plc
10 Fleet Place, Limeburner Lane, London EC4M 7SB, UK
Tel: +44 20 7575 0000
Fax: +44 20 7575 0010
Email: info@btgplc.com
Web: www.btgplc.com
CEO: Ian Harvey

Business Systems Engineering Inc
401 North Wabash, Suite 524, Chicago, IL 60611, USA
Tel: +1 312 644 1777
Fax: +1 312 644 1778
Email: bsp@bse-inc.com
Web: www.bse-inc.com
CEO: Nathan Paige

C

Cansec Systems Ltd
3105 Unity Drive, Unit 9, Mississauga, Ontario, Canada L5L 4L2
Tel: +1 905 820 2404
Fax: +1 905 820 0301
Email: sales@cansec.com
Web: www.cansec.com
Director, Sales & Marketing: Tony Williams

Cassini Advanced Systems SC
ul. Pulkowa 28, Warsaw 01-969, Poland
Tel: +48 22 835 57 57
Fax: +48 22 834 52 50
Email: info@cassini.pl
Web: www.cassini.pl

Cavio Corp
5th Floor, 21 Water St, Vancouver, British Columbia, Canada V6B 1A1
Tel: +1 604 696 0633
Fax: +1 604 696 0634
Email: info@cavio.com
Web: www.cavio.com
CEO: Paul Mann

Cecrop
Sannaedle Building 649-10, Yoksam-Dong, Kangnam-Ku, Seoul, South Korea
Tel: +82 2 222 2700
Fax: +82 2 501 6890
Email: youngkwon@cecrop.com
Web: www.cecrop.com
CEO: Lee Ki Deak

Cherry Corporation
3600 Sunset Avenue, Waukegan, IL 60087, USA
Tel: +1 847 662 9200
Fax: +1 847 360 3508
Email: cbergin@cherrycorp.com
Web: www.cherrycorp.com
President: Robert G. Terwall

Clinisync
1815 Collingswood Court, Westlake Village, CA 91362, USA
Tel: +1 805 370 3246
Fax: +1 801 749 1427
Email: Sales@clinisync.com
Web: www.clinisync.com
President: Chris Carr

Cogent Systems Inc
209 Fair Oaks Avenue, South Pasadena, CA 91030, USA
Tel: +1 626 799 8090
Fax: +1 626 799 8996
Email: info@cogentsystems.com
Web: www.cogentsystems.com

Cognitec Systems GmbH
An der Flutrinne 12, D-01139 Dresden, Germany
Tel: +49 351 86292 0
Fax: +49 351 86292 10
Email: info@cognitec-systems.de
Web: www.cognitec-systems.de
Director Sales & Marketing: Dr Jürgen Pampus

Communication Intelligence Corp
275 Shoreline Drive, Suite 500, Redwood Shores, CA 94065-1413, USA
Tel: +1 650 802 7888
Fax: +1 650 802 7777
Email: sales@cic.com
Web: www.cic.com
Chairman, President & CEO: Guido DiGregorio

CommetiX Computer Systems Inc
2872 Bristol Circle, Suite 100, Oakville, Ontario, Canada L6H 6G4
Tel: +1 905 829 9988
Fax: +1 905 829 1944
Email: eescubedo@comnetix.com
Web: www.comnetix.com
President & CEO: Edward Escubedo

CompuBlox
205 Richmond St. West. 5th Floor, Toronto, Ontario, Canada M5V 1V3
Tel: +1 416 203 8780
Fax: +1 416 203 9750
Email: info@compublox.com
Web: www.compublox.com
Sales Manager: Ron Rusnak

ComSec Enterprises
Level 3, 303 Adelaide St, Brisbane, Queensland 4000, Australia
Tel: +61 7 3222 6844
Fax: +61 7 3221 2241
Email: info@comsecent.com.au
Web: www.comsecent.com.au
Sales: Peter Lee

Configate Ltd
Rabin Science Park, Rehovot, Israel
Tel: +1 734 6473050
Fax: +1 734 6473050
Email: info@configate.com
Web: www.configate.com
Contact: Tamir Melamed

Control for Doors Ltd
Hurst Place, Woldingham Road, Woldingham, Surrey CR3 7LT, UK
Tel: +44 1883 652652
Fax: +44 1883 652055
Email: sales@cfdltd.com
Web: www.cfdltd.com

Control Module Inc
227 Brainard Road, Enfield, CT 06082, USA
Tel: +1 860 745 2433
Fax: +1 860 741 6064
Email: Sales@ControlMod.com
Web: www.controlmod.com
President: Jimmy Bianco

Cross Match Technologies Inc
3960 RCA Boulevard, Suite 6001, Palm Beach Gardens, FL 33410, USA
Tel: +1 561 622 1650
Fax: +1 561 622 1650
Email: info@crossmatch.net
Web: www.crossmatch.net
Chairman & CEO: Theodore Johnson

CSIRO
PO Box 1710, Cnr Vimiera & Pembroke Roads, Marsfield, Epping, NSW 2122, Australia
Tel: +61 2 9372 4222
Fax: +61 2 9372 4400
Email: info@tip.csiro.au
Web: www.tip.csiro.au
Division Chief: Dr Warren King

C-VIS Computer Vision und Automation GmbH
Universitätsstrasse 142, 44799 Bochum, Germany
Tel: +49 234 97066 0
Fax: +49 234 97066 30
Email: info@c-vis.com
Web: www.c-vis.com
Contact: Dr-Ing Thomas Zielke

Cyber-Card
3 Hazelcourt, 160 Witchhazel Street, Highveld Technopark, Centurion, Pretoria, South Africa
Tel: +27 12 665 0441
Fax: +27 12 665 3204
Email: info@cybergroup.co.za
Web: http://home.intekom.com/cyberbex/index.htm

Cyber-SIGN Inc
180 Montgomery St, Suite 925, San Francisco, CA 94104, USA
Tel: +1 408 324 1001
Fax: +1 408 324 1057
Email: info@cybersign.com
Web: www.cybersign.com
CEO: Daniel Dingley

D

Daon
345 Park Avenue, 17th Floor, New York, NY 10154-0037, USA
Tel: +1 212 371 2002
Fax: +1 212 371 6398
Web: www.daon.com
CEO: Oliver Tattan

Data Management Inc
7207 Wooster Pike #256, Cincinnati, OH 45227, USA
Tel: +1 859 572 0808
Fax: +1 859 572 0516
Email: sales@4dmi.com
Web: www.4dmi.com
President: Eric Mischell

Datacard Group
11111 Bren Road West, Minnetonka, MN 55343-9015, USA
Tel: +1 952 933 1223
Fax: +1 952 931 0418
Email: info@datacard.com
Web: www.datacard.com
President & CEO: Jerry E. Johnson

Datakey Electronics Inc
403 West Travelers Trail, Burnsville, MN 55337-2558, USA
Tel: +1 952 746 4066
Fax: +1 952 746 4061
Email: info@datakeyelectronics.com
Web: www.datakeyelectronics.com
President: Eric Jennings

Datastrip Ltd
7 Thame Park Business Centre, Wenman Road, Thame, Oxfordshire OX9 3XA, UK
Tel: +44 1844 215668
Fax: +44 1844 215669
Email: uk@datastrip.com
Web: www.datastrip.com
Business Development Director: Patrick Gilmore

DataTreasury Corp
175 Pinelawn Road, Suite 200, Melville, New York, NY 11747, USA
Tel: +1 631 486 5500
Fax: +1 631 486 5555
Email: info@datatreasury.com
Web: www.datatreasury.com/biometric_overview.htm
CEO: Claudio Ballard

DBA Systems
PO Box 550, 1200 S Woody Burke Rd, Melbourne, FL 32902-0550, USA
Tel: +1 321 727 0660
Fax: +1 321 952 1689
Web: www.titan.com/dbasystems/dislaw.htm

Dectel Security
Swinborne Rd. Burnt Mills, Basildon, Essex SS13 1EF, UK
Tel: +44 1268 727 586
Fax: +44 1268 591 422
Email: sales@dectel.co.uk
Web: www.dectel.co.uk
Account Manager: Chris Matthews

DelSecur Inc
1801 McGill-College, Suite 1330, Montreal, Canada H3A 2N4
Tel: +1 514 282 9000
Fax: +1 514 282 1186
Email: info@delsecur.com
Web: www.delsecur.com
Chairman, President & CEO: Pierre de Lanauze

Delsy Electronic Components AG
Am Bahnhof 2, 55765 Birkenfeld, Germany
Tel: +49 6782 9846 0
Fax: +49 6782 9846 22
Email: mail@delsy.de
Web: www.delsy.de
CEO: Dr Alfred Simmet

Dermalog GmbH
Mittelweg 120, 20148 Hamburg, Germany
Tel: +49 40 41 32 27 0
Fax: +49 40 41 32 27 41
Email: info@dermalog.de
Web: www.dermalog.de
CEO: Gunther Mull

Dialog Communication Systems Inc
13D 207 Hun-Twa S Rd. Sec 2, Taipei 10673, Taiwan
Tel: +886 2 2378 5508
Fax: +886 2 2378 3191
Email: info.asia@bioid.com.tw
Web: www.bioid.com.tw
CEO & President: Ho Chang

Digicom SpA
via A Volta 39, Cardano al Campo, VA 21010, Italy
Tel: +39 0331 702646
Fax: +39 0331 263733
Email: international@digicom.it
Web: www.digicom.it
International Account Manager: Alexander Thorak

Digimarc ID Systems
201 Burlington Road, Bedford, MA 01730, USA
Tel: +1 781 386 5710
Fax: +1 781 386 5977
Email: pdaley@digimarc.com
Web: www.digimarc-id.com
Marketing Director: Paula Daley

Digital Descriptor Systems Inc
446 Lincoln Highway, Fairless Hills, PA 19030, USA
Tel: +1 267 580 1075
Fax: +1 267 580 1090
Email: ddsihome@ddsi-cpc.com
Web: www.ddsi-cpc.com
CEO: Bob Gowell

Digital Persona Inc
805 Veterans Boulevard, Suite 301, Redwood City, CA 94063, USA
Tel: +1 650 261 6070
Fax: +1 650 261 6079
Email: info@digitalpersona.com
Web: www.digitalpersona.com
President & CEO: Fabio Righi

Digitus sro
Zerotínovo námestí 14, Prerov 750 02, Czech Republic
Tel: +420 641 219 800
Fax: +420 601 506 382
Email: info@digitus.cz
Web: www.digitus.cz
Manager: Zbynek Hanecka

Domain Dynamics Ltd
Thames Tower, Station Road, Reading, Berkshire RG1 1LX, UK
Tel: +44 118 902 3777
Fax: +44 118 902 3789
Email: info@ddl.co.uk
Web: www.ddl.co.uk
CEO: Fredrik Verkroost

E

East Shore Technologies Inc
80 Campbell Ave, Troy, New York, NY 12180, USA
Tel: +1 518 274 6510
Email: information@east-shore.com
Web: www.east-shore.com

e-Com
Level 2-147-149 Colin Street, West Perth 6005, Australia
Tel: +61 894 861455
Fax: +61 894 861499
Email: Sales@ecomaus.com
Web: www.ecomaus.com
General Manager: Michael Ryan

eCryp Inc
4712 Admiralty Way, #574, Marina Del Rey, CA 90292, USA
Email: info@ecrypinc.com
Web: www.ecrypinc.com
CEO & President: George Stickel

Ensigma Technologies
Turing House, Station Road, Chepstow,
Gloucestershire NP16 5PB, UK
Tel: +44 1291 625422
Fax: +44 1291 620301
Email: licensing@ensigma.com
Web: www.ensigma.com
VP Business Development: David McBrien

Ethentica Inc
22362 Gilberto #130, Rancho Santa Margarita,
CA 92688, USA
Tel: +1 949 858 7525
Fax: +1 949 858 7092
Email: sales@ethentica.com
Web: www.ethentica.com
President & CEO: Mark O'Hare

Evive Corporation
3 Bethesda Metro Center, Suite 700, Bethesda,
MD 20814, USA
Tel: +1 301 986 0999
Fax: +1 202 478 0441
Email: info@evive.com
Web: www.evive.com
CEO: Jonathan Eisner

Ex Cle SA
Viamonte 1653 PB, Buenos Aires 1055, Argentina
Tel: +54 1 14815 3488
Fax: +54 1 14815 3488
Email: info@ex-cle.com
Web: www.ex-cle.com

Exact Identification Corp
4650 Northgate Blvd, Suite 160, Sacramento,
CA 95834, USA
Tel: +1 916 646 4444
Fax: +1 916 646 4445
Email: steve@prosalesreps.com
Web: www.aprint.com
President & CEO: John Hunepohl

Excel Systems Inc
1F 60 Chien-Hwa 3rd St, TaoYuan 330, Taiwan
Tel: +886 3 367 7727
Fax: +886 3 367 8719
Email: mail@excelsi.com.tw
Web: www.vkeye.com

ExResource Inc
Naperville, IL 60540, USA
Tel: +1 630 573 7777
Fax: +1 630 753 0627
Email: info@exresource.com
Web: www.exresource.com

EyeDentify Europe NV
Autolei 290, Wommelgem B-2160, Belgium
Tel: +32 3 320 90 32
Fax: +32 3 320 90 33
Email: sales@eye-dentify.com
Web: www.eye-dentify.com
President: Paul De Grove

EyeNetWatch.com
Unit 2 Eelmoor Road, Eelmoor Road Ind Est,
Farnborough, Hampshire GU14 7QN, UK
Tel: +44 1252 372292
Fax: +44 1252 378167
Email: support@romseyassoc.com
Web: www.eyenetwatch.com
Contact: Mark Yadegar

EyeTicket Corp
8201 Greensboro Drive, Suite 713, McLean,
VA 22102, USA
Tel: +1 703 827 5700
Fax: +1 703 827 5705
Email: info@eyeticket.com
Web: www.eyeticket.com
CEO: Stewart Mann

F

FACE Technologies (Pty) Ltd
PO Box 25558, 263A West Avenue, 2nd Floor Meerlus
Building, Centurion, Pretoria 0105, South Africa
Tel: +27 12 633 7522
Fax: +27 12 663 7514
Email: sales@face.co.za
Web: www.face.co.za
Gen Manager: Serfies Serfontain

FaceBase Inc
711 S. Carson Street – Suite 4, Carson City, NV
89701-5217, USA
Email: ceo@facebase.com
Web: www.facebase.com

Fidelica Microsystems
423 Dixon Landing Road, Milpitas, CA 95035, USA
Tel: +1 408 941 7900
Fax: +1 408 941 8149
Email: sales@fidelica.com
Web: www.fidelica.com
VP Marketing & Business Development: Bob Allen

Financial Information Systems
PO Box 48076, Atlanta, GA 30362-1076, USA
Tel: +1 770 457 8281
Fax: +1 770 564 5548
Email: info@stopfraud.com
Web: www.stopfraud.com

Fingerprint Cards AB
PO Box 2412, Gothenburg S-403 16, Sweden
Tel: +46 31 607820
Fax: +46 31 137385
Web: www.fingerprint.se
President: Lennart Carlson

Fingerprint Technologies
Australia
Web: www.fingerprint.com.au

FingerPrint USA
PO Box 1190, Haymarket, VA 20168-8190, USA
Tel: +1 703 754 0308
Fax: +1 703 754 2251
Email: gdechman@fpusa.com
Web: www.fpusa.com
President: Gordon Dechman

FingerSec Corp
1111 Lincoln Road, Miami Beach, FL 33139, USA
Tel: +1 305 672 9910
Fax: +1 305 672 9945
Email: sales@fingersec.com
Web: www.fingersec.com

FIST
4F DaeWon Building, 694-29, Yeoksam-Dong, Kangnam-Ku, Seoul 135-081, South Korea
Tel: +82 2 539 4456
Fax: +82 2 3288 9256
Email: fist@fist.co.kr
Web: www.fist.co.kr
President: S W Cheon

Florentis Ltd
Vallis House, 57 Vallis Road, Frome, Somerset BA11 3EG, UK
Tel: +44 1373 473573
Fax: +44 1373 469850
Email: info@florentis.com
Web: www.florentis.com
Office Manager: Laraine Chalmers

Fraunhofer Institute for Integrated Circuits
Am Wolfsmantel 33, 91058 Erlangen, Germany
Tel: +49 91 317 760
Fax: +49 91 317 769 99
Email: info@iis.fhg.de
Web: www.iis.fhg.de
Head of Institute: Prof Heinz Gerhäuser PhD

Fusion Tech
B-1/105 Prerana Society, Shimpoli Rd, Shimpoli, BORIVALI (WEST), Mumbai 400 092, India
Tel: +91 22 8983126
Fax: +91 22 8983126
Email: fusiontech_india@yahoo.co.in
Web: http://web1.mtnl.net.in/~bdgala/

G

Geometrix Inc
1590 The Alameda, Suite 200, San Jose, CA 95126, USA
Tel: +1 408 999 7499
Fax: +1 408 289 8258
Email: info@geometrix.com
Web: www.geometrix.com
Chairman & CEO: David West

GEZ Microsystems Inc
3235 Shoreline Drive, Oakville, Ontario, Canada L6L 5ZI
Tel: +1 905 469 0377
Fax: +1 905 469 0327
Email: postmaster@gezmicro.com
Web: www.gezmicro.com

Graphco Technologies Inc
41 University Drive, Suite 205, Newtown, PA 18940, USA
Tel: +1 215 497 9170
Fax: +1 215 497 9321
Email: info@graphcotech.com
Web: www.graphcotech.com
Chairman & CEO: Cristian Ivanescu

Green Bit srl
Via Palmieri, 28, Turin 10138, Italy
Tel: +39 011 4342954
Fax: +39 011 4346954
Email: info@greenbit.com
Web: www.greenbit.com
President, CEO & Director: Adriano De Luca

Griaule Fingerprint Recognition
UNICAMP Technology Center, Rua Bernardo Sayão, 100, Cidade Universitária, Campinas – SP 13084-971, Brazil
Tel/Fax: +55 19 3788 4998
Email: info@griaule.com
Web: www.griaule.com
President: Iron Daher

Guardware Systems Ltd
Üllõi út 102, Budapest 1089, Hungary
Tel: +36 1 323 2500
Fax: +36 1 323 2501
Email: info@guardware.com
Web: www.guardware.com
Sales & Marketing Manager: Tamás Kramer

H

Hectrix Ltd
Unit 002, Tech Centre, 72 Tat Chee Avenue, Kowloon Tong, Hong Kong
Tel: +852 2319 1333
Fax: +852 2776 8997
Email: info@hectrix.com
Web: www.hectrix.com
Chairman & CEO: Thomas Wan

Heimann Biometric Systems GmbH
Unstrutweg 4, D-07743 Jena, Germany
Tel: +49 3641 429713
Fax: +49 3641 429714
Email: info@hbs-jena.com
Web: www.hbs-jena.com
CEO & President: Dr Bernd Reinhold

Hunno Technologies Inc
705-18, 12F, Sewon Venture Town, Yoksam-Dong, Kangnam-Ku, Seoul 135-080, South Korea
Tel: +82 2 2008 0800
Fax: +82 2 2008 0805
Email: info@hunno.com
Web: www.hunno.com
Chief Manager: Andy Kim

Hyundai Information Technology Co Ltd
60-1 Kukdong Bldg, Chungmuro 3 Ka, Jung Ku, Seoul 100-705, South Korea
Tel: +82 2 2129 4114
Email: eclipse@hit.co.kr
Web: www.hit.co.kr
CEO: Sun-Bae Kim

I

I/O Software Inc
1533 Spruce Street, Riverside, CA 92507, USA
Tel: +1 909 222 7600
Fax: +1 909 222 7601
Email: info@iosoftware.com
Web: www.iosoftware.com
CEO: William Saito

IBIA
601 Thirteenth Street, NW, Suite 370 South, Washington DC 20005, USA
Tel: +1 202 783 7272
Fax: +1 202 783 4345
Email: ibia@ibia.org
Web: www.ibia.org
Executive Director: Richard Norton

IBMC Inc
PO Box 4179, Clearwater, FL 33758-4179, USA
Tel: +1 727 507 0506
Email: fduss@ibmc.com
Web: www.ibmc.com
Contact: Frank A Duss

ID Technologies Corp
2506 W. Nash St, Suite C, Wilson, NC 27896, USA
Tel: +1 252 206 1089
Fax: +1 252 206 4990
Email: information@idtek.com
Web: www.idtek.com
Chairman: William F Lane

Ideal Software Corp
PO Box 182756, 157 Zahran Street, Amman 11118, Jordan
Tel: +962 6 593 0123
Fax: +962 6 593 0388
Email: info@idealsoft.com.jo
Web: www.idealsoft.com.jo
Chairman: Karim Kawar

Idencom AG
Im staubeweidli 11, CH-8603 Wädenswil, Switzerland
Tel: +41 1 780 66 44
Fax: +41 1 780 66 46
Email: info@idencom.com
Web: www.idencom.com

IdentAlink
34/36 High Street, Kettering, Northamptonshire NN16 8SU, UK
Tel: +44 1536 483000
Fax: +44 1536 483817
Email: info@biometrics.ws
Web: www.identalink.com

IDentification Systems Inc
1125 W. Vine Street, Stockton, CA 95203, USA
Tel: +1 209 937 0833
Fax: +1 209 937 0834
Email: info@idyou.com
Web: www.idyou.com

Identix Inc
5600 Rowland Road, Minnetonka, MN 55343, USA
Tel: +1 952 932 0888
Fax: +1 952 932 7181
Email: sales@identix.com
Web: www.identix.com
President & CEO: Joseph J Atick PhD

Idex ASA
PO Box 519, Gamle Borgenvei 5, Asker N-1373, Norway
Tel: +47 66 98 77 00
Fax: +47 66 98 77 11
Email: mailbox@idex.no
Web: www.idex.no
CEO: Andrew Heap

IDIAP
PO Box 592, Rue du Simplon 4, CH-1920 Martigny, Switzerland
Tel: +41 27 721 77 11
Fax: +41 27 721 77 12
Email: secretariat@idiap.ch
Web: www.idiap.ch

IDTECK Co Ltd
5F, Ace Techno Tower B/D, Deungchon-Dong, Kangsuh-Gu, Seoul 157-030, South Korea
Tel: +82 2 659 0055
Fax: +82 2 659 0086
Email: mikeyoo@idteck.com
Web: www.idteck.com
Vice President: Mike Yoo

IDynta Systems Inc
7 South Prospect Ave, Clarendon Hills, IL 60514, USA
Tel: +1 630 789 2777
Email: webmaster@idynta.com
Web: www.idynta.com
Contact: Tom Duffy

iGuard Security Systems
See Hectrix above
Hong Kong
Email: info@hectrix.com
Web: www.iguardsystem.com

ILI Technologies Limited
11, BTM Ring Road, BTM Layout 2nd Stage, Bangalore, Karnataka 560076, India
Tel: +91 80 678 9569
Fax: +91 80 678 9635
Email: info@ilitec.com
Web: www.ilitec.com
Technical Executive: M A Srinivasa Rao

Image Metrics plc
Regent House, Heaton Lane, Stockport, Cheshire SK4 1BS, UK
Tel: +44 161 476 8220
Fax: +44 161 480 4583
Email: info@image-metrics.com
Web: www.image-metrics.com
CEO: David Pilgrim

ImageWare Systems Inc
10883 Thornmint Road, San Diego, CA 92127, USA
Tel: +1 858 673 8600
Fax: +1 858 673 1770
Email: sales@iwsinc.com
Web: www.iwsinc.com
President & CEO: S James Miller Jr

Imaging Automation Inc
25 Constitution Drive, Bedford, NH 03110-6000, USA
Tel: +1 603 471 9325
Fax: +1 603 471 9326
Email: info@imagingauto.com
Web: www.imagauto.com
CEO: Bill Thalheimer

Imagis Technologies Inc
Suite 1630, 1075 West Georgia Street, Vancouver, British Columbia, Canada V6E 3C9
Tel: +1 604 684 2449
Fax: +1 604 684 9314
Email: info@imagistechnologies.com
Web: www.imagistechnologies.com
President & CEO: Iain Drummond

Inception Technologies Inc
Bellflower, California, USA
Tel: +1 800 473 9457
Fax: +1 562 925 9655
Email: sales@inceptiontechnologies.com
Web: www.inceptiontechnologies.com
Sales Manager: Lowell Schwartz

Indivos Corporation
155 Grand Avenue, Suite 1050, Oakland, CA 94612, USA
Tel: +1 510 903 1000
Fax: +1 510 268 0935
Email: info@indivos.com
Web: www.indivos.com
Senior Vice President, Marketing & Business Development: Frank S Pierce

Infineon Technologies AG
St.-Martin-Str 53, 81669 Munich, Germany
Tel: +49 89 234 00
Fax: +49 89 234 2 46 94
Email: company.info@infineon.com
Web: www.infineon.com
President & CEO: Dr Ulrich Schumacher

Ingenico
9, Quai de Dion Bouton, Puteaux Cedex 92816, France
Tel: +33 1 46 25 82 00
Fax: +33 1 47 72 56 95
Email: jlfilippini@ingenico.fr
Web: www.ingenico.fr
Chairman: J J Poutrel

Inroad Solutions
127-116 Research Drive, Saskatoon, Canada S7N 3R3
Tel: +1 306 931 4556
Fax: +1 306 931 4553
Email: info@inroad.net
Web: www.inroad.ca/biometrics.html
Director: Tony Bassett

Integris (now Steria)
Great West Road, Brentford, Middlesex TW8 9DH, UK
Tel: +44 870 600 4466
Fax: +44 1442 885122
Email: information@steria.co.uk
Web: www.steria.co.uk
CEO: John Torrie

Interactive Products Inc
55A Oakway Center, Eugene, OR 97401, USA
Tel: +1 541 341 4964
Fax: +1 541 341 4965
Email: sales@ipispeech.com
Web: www.ipispeech.com
President & CEO: Hugh Boyette

International Biometric Group
One Battery Park Plaza, New York, NY 10004, USA
Tel: +1 212 809 9491
Fax: +1 212 809 6197
Email: contact@biometricgroup.com
Web: www.biometricgroup.com
Director of Special Projects: Michael Thieme

International Electronics Inc
427 Turnpike Street, Canton, MA 02021, USA
Tel: +1 781 821 5566
Fax: +1 781 821 4443
Email: johnw@ieib.com
Web: www.ieib.com
President & CEO: John Waldstein

Intervid International
PO Box 1218, 6 Charles Way, Kloof, Durban 3640, South Africa
Tel: +27 31 764 9700
Fax: +27 31 764 2558
Email: mail@intervid.com
Web: www.intervid.com
Managing Director: Rob Le Seuer

Intervoice Inc
17811 Waterview Parkway, Dallas, TX 75252, USA
Tel: +1 972 454 8000
Fax: +1 972 454 8676
Email: simon.edwards@uk.intervoice.com
Web: www.intervoice.com
CEO & Chairman: David W Brandenburg

Intuate Biometrics SL
C/ Numancia 43-45, Entlo. 3ª, Barcelona 08029, Spain
Tel: +34 93 162 4355
Fax: +34 93 363 2881
Email: info@intuate.com
Web: www.intuate.com
Director: Sergio Peirot

IQ Biometrix Inc
1630 E. Shaw Ave, Suite 192, Fresno, CA 93710, USA
Tel: +1 559 222 2229
Fax: +1 559 222 2206
Email: info@iqbiometrix.com
Web: www.iqbiometrix.com
CEO: Richard K Thompson

IQS Biometric Solutions
PO Box 2109, Kiryat Byalik 27120, Israel
Tel: +972 4 8400303
Fax: +972 4 8400539
Email: info@biometric-center.com
Web: www.biometric-center.com
CEO: Shalev Baruch

Iridian Technologies Inc
121 Whittendale Drive, Suite B, Moorestown, NJ 08057, USA
Tel: +1 856 222 9090
Email: iridian@iridiantech.com
Web: www.iridiantech.com
President & CEO: Bill Voltmer

J

J. Markowitz Consultants
5801 North Sheridan Road, 19A, Chicago, IL 60660, USA
Tel: +1 773 769 9243
Fax: +1 773 769 9253
Email: info@jmarkowitz.com
Web: www.jmarkowitz.com
President: J Markowitz

Jaypeetex Engineering Pvt Ltd
403 Bhagyalaxmi, Kennedy Bridge, Opera House, Mumbai 400 004, India
Tel: +91 22 387 6662
Fax: +91 22 387 6663
Email: jaypeetex@vsnl.com
Web: www.jaypeetex.com
Chairman: BK Podar

K

Keico Hi Tech Inc
ACE TechnoTower 9th floor 197-17, GURO-3DONG GURO-GU, Seoul 152-053, South Korea
Tel: +82 2 853 9000
Fax: +82 2 864 8800
Email: keico@keico.co.kr
Web: www.keico.co.kr

KeyTronicEMS Co
North 4424 Sullivan Road, Spokane, WA 99214, USA
Tel: +1 509 928 8000
Fax: +1 509 927 5383
Email: marketing@keytronic.com
Web: www.keytronic.com
President & CEO: Jack Oehlke

Keyware Technologies NV
Joseph Wybranlaan 40, Brussels B-1070, Belgium
Tel: +32 2 526 16 16
Fax: +32 2 524 38 67
Email: info@keyware.com
Web: www.keyware.com
General Manager: Chris Buyse

Kinetic Sciences Inc
655 West Kent Ave N, Unit 500, Vancouver, British Columbia, Canada V6P 6T7
Tel: +1 604 267 7805
Fax: +1 604 267 7806
Email: info@kinetic.bc.ca
Web: www.kinetic.bc.ca
President & CEO: Guy Immega

Kingshad International
2nd Floor, Dabo shopping Mall, Plot 1240, Samuel Ladoke Akintola Boulevard, Garki II, Abuja, Nigeria
Tel: +234 803 313 8145
Fax: +234 9 234 7850
Email: enquiries@kingshad.com
Web: www.kingshad.com
General Manager & CEO: Emmanuel N Michael

Konetix Inc/TimeCentre
5660 Airport Blvd, Boulder, CO 80301, USA
Tel: +1 303 444 8771
Fax: +1 303 442 6286
Email: info@konetix.com
Web: www.labortracking.com
Sales Director: Rich Buteau

L

Labcal Technologies Inc
400, boul. Jean-Lesage, Hall Ouest, Suite 30, Quebec, Canada G1K 8W1
Tel: +1 418 692 3137
Fax: +1 418 692 1488
Email: info@labcal.com
Web: www.labcal.com
President: Carl Boudreau

Lockheed Martin Information Technology
7375 Executive Plaza, Seabrook, MD 20706, USA
Tel: +1 301 352 2650
Web: www.it.lockheedmartin.com
President: Linda Gooden

Lone Wolf Software Inc
10054 E. Alondra, Suite 201, Bellflower, CA 90706, USA
Tel: +1 800 473 9457
Fax: +1 562 925 9655
Email: sales@lonewolfsoftware.com
Web: www.timewolf.com

Loquendo SpA
Via Nole 55, Turin 10149, Italy
Tel: +39 011 757 6111
Fax: +39 011 757 6909
Email: info@loquendo.com
Web: www.loquendo.com
Managing Director & CEO: Silvano Giorcelli

Lumidigm Inc
800 Bradbury SE, Suite 213, Albuquerque, NM 87106, USA
Tel: +1 505 272 7406
Fax: +1 505 272 7354
Email: lumidigminfo@lumidigm.com
Web: www.lumidigm.com
Contact: Rob Rowe

M

Mason Vactron Ltd
25 Swan Lane, Evesham, Worcestershire WR11 4PE, UK
Tel: +44 1386 765397
Fax: +44 1386 765351
Email: sales@masonvactron.co.uk
Web: www.masonvactron.co.uk
Managing Director: Bob Freeman

MESA srl
Loc Indicatore 60/g, Arezzo 52100, Italy
Tel: +39 0575 987080
Fax: +39 0575 929254
Web: www.mesa-sic.com

Miaxis Biometrics Co Ltd
3/F Building 1, East Software Park, 90 Wensan Rd, Hangzhou 310012, China
Tel: +86 571 88473588
Fax: +86 571 88473523
Email: business@miaxis.com
Web: www.miaxis.com
General Manager: Li Jian

Microsoft Corporation
One Microsoft Way, Redmond, WA 98052 6399, USA
Tel: +1 425 882 8080
Web: www.microsoft.com
CEO: Steve Ballmer

MotionTouch Ltd
PSC House, Jessamy Road, Weybridge, Surrey KT13 8LD, UK
Tel: +44 1932 854721
Fax: +44 1932 854940
Email: info@motiontouch.com
Web: www.motiontouch.com
Sales Director: Henry Powell

Motorola Semiconductor Products
505 Barton Springs Road, Suite 400, Austin, TX 78704, USA
Tel: +1 512 505 8801
Fax: +1 512 505 8811
Web: www.motorola.com
Vice President & General Manager SPS: Pete Shinyeda

N

National Biometric Test Center
San Jose State University, 1 Washington Square, San Jose, CA 95192, USA
Tel: +1 408 924 4037
Fax: +1 408 924 3883
Email: biomet@email.sjsu.edu
Web: www.engr.sjsu.edu/biometrics
Director: Dr Jim Wayman

Navigator Solutions Ltd
Navigator House, Pleckgate Road, Lancashire BB1 8QW, UK
Tel: +44 1254 244530
Fax: +44 1254 244531
Email: fingertech@navigator-solutions.co.uk
Web: www.navigator-solutions.co.uk

NCT Group Inc
20 Ketchum Street, Westport, CT 06880, USA
Tel: +1 203 226 4447
Fax: +1 203 226 3123
Email: sales@nctgroupinc.com
Web: www.nct-active.com
Chairman & CEO: Michael Parrella

NEC Technologies
1250 N. Arlington Heights Road, Itasca, IL 60143-1248, USA
Tel: +1 630 467 5000
Fax: +1 630 467 4363
Email: salesinfo@nectech.com
Web: www.nectech.com

Net Nanny Software Inc
15831 NE 8th Street, Suite 200, Bellevue, WA 98008, USA
Tel: +1 425 688 3008
Fax: +1 425 688 3010
Email: sales@netnanny.com
Web: www.netnanny.com
Chairman, CEO & CTO: Gordon Ross

Neuricam SpA
Via S Maria Maddalena 12, Trento 38100, Italy
Tel: +39 0461 260552
Fax: +39 0461 260617
Email: info@neuricam.com
Web: www.neuricam.com
CEO: Dr Alvise Sartori

Neurodynamics Ltd
Cambridge Business Park, Cowley Road, Cambridge, Cambridgeshire CB4 0WZ, UK
Tel: +44 1223 488540
Fax: +44 1223 488541
Email: info@neurodynamics.com
Web: www.neurodynamics.com
Contact: Dr Mike Lynch

Neurotechnoligija Ltd
Ateities str 10, Vilnius 2057, Lithuania
Tel: +370 277 33 15
Fax: +370 277 33 16
Email: neurotec@neurotechnologija.com
Web: www.neurotechnologija.com

Neusciences
Unit 2, Lulworth Business Centre, Nutwood Way, Totton, Southampton, Hampshire SO40 3WW, UK
Tel: +44 23 8066 4011
Fax: +44 23 8087 3707
Email: biometrics@neusciences.com
Web: www.neusciences.com/biometrics
General Manager: Brian Kett

Nextern Co Ltd
2-3F, Jieun Bldg, 374-5 Seokyo-dong, Mapho-ku, Seoul 121-839, South Korea
Tel: +82 2 330 8585
Fax: +82 2 330 8599
Email: webmaster@nextern.com
Web: www.nextern.com
CEO: Andrew Kim

Nexus Group International Inc
399 Pearl Street, Burlington, Ontario, Canada L7R 2M8
Tel: +1 905 634 4111
Fax: +1 905 634 1101
Email: info@nxsgrp.com
Web: www.nxsgrp.com
CEO: David Lobb

NIST
100 Bureau Drive, Stop 4700, Gaithersburg, MD 20899-4700, USA
Tel: +1 301 975 3543
Fax: +1 301 926 9524
Email: atp@nist.gov
Web: www.atp.nist.gov
Contact: Dr Omid M Omidvar

NITGen Co Ltd
Korea Sanhak Foundation Bldg 5F, Seocho-dong, Seocho-gu, Seoul 1337-31, South Korea
Tel: +82 2 3415 1800
Fax: +82 2 3415 1801
Web: www.nitgen.com/eng
President: Jun Young Ahn

Northrop Grumman Information Technology TASC
4801 Stonecroft Boulevard, Chantilly, VA 20151-3822, USA
Tel: +1 703 633 8300
Fax: +1 703 449 3400
Email: info@tasc.com
Web: www.tasc.com
President: Jim O'Neill

Novell (NMAS)
1800 South Novell Place, Provo, UT 84606-6194, USA
Tel: +1 888 321 4272
Email: crc@novell.com
Web: www.nxssg.net
CEO: Jack L Messman

NOVUS Technology Inc (formerly LARES)
1505 E. Houston St, San Antonio, TX 78202, USA
Tel: +1 210 223 4377
Fax: +1 210 354 4075
Email: info@novus-security.com
Web: www.novussecurity.com
Chairman: Michael Cation

Nuance Communications Inc
1005 Hamilton Court, Menlo Park, CA 94025, USA
Tel: +1 650 847 0000
Fax: +1 650 847 7979
Email: info@nuance.com
Web: www.nuance.com
President & CEO: Ronald Croen

O

Oki Electric Industry Co Ltd
7-12, Toranomon 1-chome, Minato-ku, Tokyo 105-8460, Japan
Tel: +81 3 3501 3111
Fax: +81 3 3581 5522
Web: www.oki.com/jp
General Manager: Haruo Kimura

Omron Technology Ventures Group
3945 Freedom Circle, Suite 1070, Santa Clara, CA 95054, USA
Tel: +1 408 970 1190
Fax: +1 408 727 5540
Email: jmcnamara@itc-omron.com
Web: www.society.omron.com/faceid/index.html
Vice President: Jerry McNamara

Optel Ltd
Otwarta 10a, Warsaw PL-50-212, Poland
Tel: +48 71 329 68 53
Fax: +48 71 329 68 52
Email: optel@optel.com.pl
Web: www.optel.com.pl

Optical Recognition Objectives Ltd
Leningradskaya 54/30, Samara 443099, Russia
Tel: +7 1561 264 2371
Fax: +7 1561 264 2371
Email: sales@gabitus.com
Web: www.gabitus.com
CEO: Alexander Filatov

Opticom Technologies Ltd
Shalom Tower, 9 Ahad Ha'am Street, Tel Aviv, Israel
Tel: +972 3 5100170
Fax: +972 35100171
Email: opticomb@inter.net.il
Web: www.opticom-tech.com

ORGA Kartensysteme GmbH
Am Hoppenhof 33, 33104 Paderborn, Germany
Tel: +49 5251 889 0
Fax: +49 5251 889 1889
Email: info@orga.com
Web: www.orga.com
CEO: Dr Lando Zappei

Oxford Micro Devices Inc (OMDI)
Lantern Ridge Office Park, 731 Main Street, Building 2, Monroe, CT 06468, USA
Tel: +1 203 445 0562
Fax: +1 203 445 0564
Email: info@omdi.com
Web: www.omdi.com
CEO: Steve Morton

P

Panasonic UK
Panasonic House, Willoughby Road, Bracknell, Berkshire RG12 8FP, UK
Tel: +44 1344 853176
Fax: +44 1344 853705
Web: www.panasonic.co.uk/nw/home/
Managing Director: H Otsuki

Persay Inc
234 Crossways Park Drive, Woodbury, New York, NY 11797, USA
Tel: +1 516 677 7291
Fax: +1 516 677 7197
Email: tom_cheifetz@persay.com
Web: www.persay.com
CEO: Almog Aley-Raz

Photobase Ltd
Photobase House, 518 Watford Way, Mill Hill, London NW7 2PT, UK
Tel: +44 20 8906 4060
Fax: +44 20 8906 3605
Email: livelink@photobase.co.uk
Web: www.photobase.co.uk
Managing Director: Jeremy Rose

Plettac Electronic Security GmbH
Wuerzburger Strasse 150, 90766 Fuerth, Germany
Tel: +49 911 758 84 495
Fax: +49 911 759 12 71
Email: info@plettac-electronics.de
Web: www.plettac-electronics.com/zks

Polaroid ID
784 Memorial Drive, Cambridge, MA 02139, USA
Tel: +1 781 386 2000
Web: www.polaroid.com
Chairman & CEO: Gary T. DiCamillo

Precise Biometrics AB
Dag Hammarskjölds v 2, Lund SE – 224 64, Sweden
Tel: +46 46 311 100
Fax: +46 46 311 101
Web: www.precisebiometrics.com
CEO: Christer Bergman

Print Electronics
A-9, Royal Industrial Estate, Naigaum X Road, Wadala, Mumbai 400 031, India
Tel: +91 22 411 1160/1/2
Fax: +91 22 413 5883
Email: peepl@vsnl.com
Web: www.printelectronics.com

Printrak International
1250 North Tustin Avenue, Anaheim, CA 92807, USA
Tel: +1 714 238 2000
Fax: +1 714 237 0018
Email: info@printrak.com
Web: www.printrakinternational.com
President: Dan Crawford

PrintScan International Inc
Wächterweg 4, CH-5707 Seengen, Switzerland
Tel: +41 62 767 7032
Fax: +41 62 767 7033
Email: ce@ceag.ch
Web: www.printscan.net
Technical Director: Dr Carsten Emde

Prism Consulting
USA
Tel: +1 703 938 8764
Email: info@prismtotalsolutions.com
Web: www.prismtotalsolutions.com

Privacy Curtain
1216 Patrick Street, Kissimmee, FL 34741, USA
Tel: +1 800 274 0545
Fax: +1 407 870 1943
Email: info@CCM2.com
Web: www.privacycurtain.com
President CC&M: Rocky Dunkman

Proglobo Biometrics Technologies
Rua Castilho 71, 2º Esq, Lisbon 1250-170, Portugal
Tel: +351 21 792 49 11
Fax: +351 21 792 49 59
Email: info@proglobo.com
Web: www.proglobo.com
CEO: David Bernabé Fernandes

Protocom Development Systems
Level 1, 7 Phipps Close, DEAKIN ACT, Canberra 2600, Australia
Tel: +61 2 6208 4888
Fax: +61 2 6281 7460
Web: www.protocom.cc
CEO: Jason Hart

Q

Qvoice Inc
13 Kilroy Rd, Newton, NJ 07860, USA
Tel: +1 973 786 6878
Fax: +1 973 786 6131
Email: QVoice@warwick.net
Web: www.qvtrek.com
President: Norman Hughes

R

Rayco Security Inc
7737 Densmore Ave, Van Nuys, CA 91406, USA
Tel: +1 818 779 0900
Fax: +1 818 779 0103
Email: info@raycosecurity.com
Web: www.raycosecurity.com

Recognition Systems Inc
1520 Dell Avenue, Campbell, CA 95008, USA
Tel: +1 408 341 4135
Fax: +1 408 341 4101
Email: RSIsales@irco.com
Web: www.recogsys.com
General Manager: Martin Huddart

Retinal Technologies Inc
1st Floor, 30 Union Wharf, Boston, MA 02109, USA
Tel: +1 617 723 6579
Fax: +1 617 720 5260
Email: info@retinaltech.com
Web: www.retinaltech.com
Contact: Paul Steitz

Ringdale UK Ltd
56 Victoria Road, Burgess Hill, West Sussex RH15 9LR, UK
Tel: +44 1444 871349
Fax: +44 1444 870228
Email: info@ringdale.com
Web: www.ringdale.com

Rofin Australia Pty Ltd
6/42-44 Garden Blvd, Dingley, Victoria, Australia
Tel: +61 3 9558 0344
Fax: +61 3 9558 0252
Email: info@rofin.com.au
Web: www.rofin.com.au
Managing Director: Garry Yost

S

SAFLINK Corporation
11911 NE 1st St, Suite B-304, Bellevue, WA 98005, USA
Tel: +1 425 278 1100
Fax: +1 425 278 1300
Email: info@saflink.com
Web: www.saflink.com
President & CEO: Glenn Argenbright

Sagem Morpho
1145 Broadway Plaza, Suite 200, Tacoma, WA 98402, USA
Tel: +1 253 383 3617
Fax: +1 253 591 8856
Email: info@morpho.com
Web: www.morpho.com
CEO: Jean-Marc Suchier

ScanSoft BVBA
Guldensporenpark 32, Building D, Merelbeke 9820, Belgium
Tel: +32 9 239 8000
Fax: +32 9 239 8001
Web: www.scansoft.com
Vice President: Wayne Crandall

SD Industries GmbH
Äussere Günzburger Strasse 2-10, 89423 Gundelfingen, Germany
Tel: +49 9073 95 89 90
Fax: +49 9073 95 89 98 0
Email: info@sd-industries.com
Web: www.sd-industries.com
Manager: Dipl-Ing Erhard Irion

SecuGen Corporation
348 Montague Expressway, Milpitas, CA 95035, USA
Tel: +1 408 942 3400
Fax: +1 408 262 0264
Email: sales@secugen.com
Web: www.secugen.com
CEO: Robert Kyle

SecureTech Solutions Inc
6542 Hypoluxo Road #137, Lake Worth, FL 33467, USA
Tel: +1 561 433 9314
Fax: +1 561 423 8397
Email: info@securetechsolutions.com
Web: www.securetechsolutions.com

Securicor IS
Securicor Information Systems Ltd, Digital Security, Methuen Park, Chippenham, Wiltshire SN14 0TW, UK
Tel: +44 7002 919999
Fax: +44 7002 939999
Email: drew.paterson@sis.securicor.co.uk
Web: www.securicor.com/sis
Manager: Drew Paterson

Security Biometrics Inc
1030 West Georgia Street, 14th Floor, Vancouver, Canada V6E 2Y3
Tel: +1 604 609 7749
Fax: +1 604 609 7715
Email: info@sigbio.com
Web: http://sigbio.xplorex.com
CEO: Kenneth P. Barr

Security Information Technology
7F West Wing, IT Venture Tower, 78 Garak-dong, Songpa-gu, Seoul 138-160, South Korea
Tel: +82 2 2142 2700
Fax: +82 2 2142 2727
Web: www.secuit.com
CEO: Joohyung Lee

Senex Technologies Co Ltd
5th Fl. Seojin Bldg. 237-10, Nonhyun-dong, Kangnam-gu, Seoul 135-010, South Korea
Tel: +82 2 2056 4341
Fax: +82 2 548 1202
Email: senextech@senextech.com
Web: www.senextech.com
President: Mr Namgoong

Sense Holdings Inc
4503 NW 103 Ave (Suite 200), Sunrise, FL 33351, USA
Tel: +1 954 726 1422
Fax: +1 954 726 2022
Email: sales@senseme.com
Web: www.senseme.com
CEO: Dore Scott Perler

Sentry Com Ltd
9 International Str, 2nd Floor, Ramot Remez, Haifa 32207, Israel
Tel: +972 4 812 2250
Fax: +972 3 725 5867
Email: rtalmor@sentry-com.co.il
Web: www.voiceprove.com
Contact: Dr Eli Talmor

SentryTechnology Inc
PO Box 1022, Bensalem, PA 19020, USA
Tel: +1 215 28 3690
Email: robert.waryga@verizon.net
Web: www.sentrykids.com
Contact: Robert Waryga

Serco Group plc
Dolphin House, Windmill Road, Sunbury-on-Thames, Middlesex TW16 7HT, UK
Tel: +44 1932 755900
Fax: +44 20 8334 4301
Email: generalenquiries@serco.com
Web: www.serco.com
Executive Chairman: Richard White

Siemens Biometrics
ICM RDC IS BIO, Hofmannstr 51, 81359 Munich, Germany
Tel: +49 89 7 22 49380
Fax: +49 89 7 22 49015
Email: biometrics@mch.siemens.de
Web: www.designbytechnology.de
Business Development Manager: Dr Markus Ludwig

Siemens PSE TechLab
3870 North First Street, San Jose, CA 95134, USA
Tel: +1 408 456 4153
Fax: +1 408 456 4312
Email: contact@psetechlab.com
Web: www.siemensidmouse.com
Regional Sales Director: Mike Weber

Sign@Tech
1 Ville-Marie Place, Suite 2001, Montreal, Quebec, Canada H3B 2C4
Tel: +1 514 874 0888 ext 247
Fax: +1 514 874 0393
Email: info@signatechsys.com
Web: www.signatechsys.com

Simple Technology Inc
175 Romina Drive, Concord, Ontario, Canada L4K 4V3
Tel: +1 905 669 0300
Fax: +1 905 669 0676
Email: sales@simpletechnology.com
Web: www.simpletechnology.com

Sintec Corporation
Bucheon-Technopark 202-504, Yakdae-dong, Wonmi-gu, Bucheon, Kyonggi, South Korea
Tel: +82 32 234 2000
Fax: +82 32 234 2020
Email: sit@sitway.com
Web: www.sitway.com
Managing Director: Francis S Kim

Smart Biometrics Inc
2772 Depot Avenue, Sanford, FL 32773, USA
Tel: +1 407 688 8266
Fax: +1 407 688 8271
Email: sales@smartbiometrics.com
Web: www.smartbiometrics.com
CEO: David R Allen

Softpro
Wilhelmstrasse 34, 71034 Boeblingen, Germany
Tel: +49 7031 66 06 0
Fax: +49 7031 66 06 66
Email: info@softpro.de
Web: www.signplus.com
Contact: Jörg-M Lenz

Software Professional GmbH & Co
Wilhelmstrasse 34, 71034 Boeblingen, Germany
Tel: +49 7031 66 06 0
Fax: +49 7031 66 06 66
Email: info@softpro.de
Web: www.signplus.com
Managing Director & CEO: Rudolf Sprengel

Sonda
Mkakeeva Ave 21, Miass City, Cheliabinsk Region 456 318, Russia
Tel: +7 35135 276 77
Fax: +7 35135 298 95
Email: sonda@sonda.ru
Web: www.sonda.ru
General Director: Anatoly Bokov

Sonetech Corp
10 Commerce Park North, Unit #1, Bedford, NH 03110-6906, USA
Tel: +1 603 626 0315
Fax: +1 603 624 8073
Web: www.sonetechcorp.com
CEO: Dr Harvey Woodsum

Speech Technology Center
4 Krasutskogo, St Petersburg 196084, Russia
Tel: +7 812 325 8848
Fax: +7 812 327 9297
Email: info@speechpro.com
Web: www.speechpro.com
President: Dr Michael V Khitrov

SpeechWorks International Inc
695 Atlantic Avenue, Boston, MA 02111, USA
Tel: +1 617 428 4444
Fax: +1 617 428 1122
Email: info@speechworks.com
Web: www.speechworks.com
CEO: Stuart Patterson

SQN Banking Systems
PO Box 423, 65 Indel Ave, Rancocas, NJ 08073, USA
Tel: +1 609 261 5500
Fax: +1 609 265 9517
Email: info@sqnbankingsystems.com
Web: www.sqnbankingsystems.com

Startek Engineering Inc
3F, 54 Park Ave. II, Science-Based Industrial Park, Hsinchu 300, Taiwan
Tel: +886 3 578 5389
Fax: +886 3 578 7089
Email: sales@mail.startek-eng.com
Web: www.startek.com.tw
General Manager: Dr Wen-Hsin Hsu

STMicroelectronics
1000 East Bell Road, Phoenix, AZ 85022, USA
Tel: +1 602 485 6100
Fax: +1 602 485 6102
Web: www.st.com
President & CEO: Pasquale Pistorio

Stromberg LLC
525 Technology Park, Suite 165, Lake Mar, FL 32746, USA
Tel: +1 407 333 3282
Fax: +1 407 333 4333
Email: info@stromberg.com
Web: www.stromberg.com
President & CEO: Ted Dinkel

Suprema Inc
Dongchun Bldg. 13-21 Yangjae dong, Seocho, Seoul 137-130, South Korea
Tel: +82 2 571 9305
Fax: +82 2 571 9306
Email: info@suprema.co.kr
Web: www.suprema.co.kr
President & CEO: James (Jae-Won) Lee

Symtron Technology Inc
2F, 129 Section 3, Ming Sheng East Road, Taipei 106, Taiwan
Tel: +886 2 2545 8218
Fax: +886 2 2545 8219
Email: feedback@symtrontech.com
Web: www.symtrontech.com
President & CEO: Dr Alan Yeh

SyntheSys Secure Technologies Inc
550 Fairway Drive, Suite 110, Deerfield Beach, FL 33441, USA
Tel: +1 954 427 7400
Fax: +1 954 480 9717
Email: information@synthesysusa.com
Web: www.synthesysusa.com
CEO: Shashi Kapur

T

Techno Imagia Co Ltd
7F B B Daiichi Bldg, 1-8-8 Ryogoku, Sumida-ku, Tokyo 130-0029, Japan
Tel: +81 3 5600 0980
Fax: +81 3 5600 0981
Email: sales@Technoimagia.co.jp
Web: www.technoimagia.co.jp

Technogama UAB
Kalvariju 125-4d, Vilnius 2042, Lithuania
Tel: +370 5 270 0048
Fax: +370 5 270 0049
Email: tg@technogama.lt
Web: www.technogama.lt
Marketing Manager: Virmantas Galdikas

TeKey Research Group
PO Box 6026, Ashqelon 78160, Israel
Tel: +972 52 477776
Fax: +972 8 6762241
Email: info@tekey.com
Web: www.tekey.com
Marketing Manager: Uri Haas

TeleTrusT Deutschland eV
Chamissostrasse 1, D-99096 Erfurt, Germany
Tel: +49 361 34 60 531
Fax: +49 361 34 53 957
Email: info@teletrust.de
Web: www.teletrust.de
Managing Director: Prof Dr Helmut Reimer

Tennyson Biometric Systems
1328 Oak Valley Drive, Mt Juliet, TN 37122, USA
Tel: +1 615 758 7460
Fax: +1 615 773 8410
Email: rspencer@tennysonsystems.com
Web: www.tennysonsystems.com
Contact: R Spencer

The International Biometric Society
1444 I Street NW, Suite 700, Washington DC 20005-6542, USA
Tel: +1 202 712 9049
Fax: +1 202 216 9646
Email: ibs@bostromdc.com
Web: www.tibs.org
President: Norman E Breslow

The Ottawa Technology Group Inc
204 – 1900 City Park Drive, Ottawa, Ontario, Canada K1J 1A3
Tel: +1 613 745 4441
Fax: +1 613 745 7473
Email: info@otg.ca
Web: www.otg.ca

The Ottawa Telephony Group Inc
204 – 1900 City Park Drive, Ottawa, Ontario, Canada K1J 1A3
Tel: +1 613 745 4441
Fax: +1 613 745 7473
Email: info@otg.ca
Web: www.otg.ca

The Phoenix Group Inc
205 N Walnut, Pittsburg, KS 66762, USA
Tel: +1 620 232 6420
Fax: +1 620 232 2606
Email: phoenix@afix.net
Web: www.afix.net

ThumbAccess Biometrics Corp
Level 20, 99 Walker Street, North Sydney, NSW 2060, Australia
Tel: +61 2 96571360
Fax: +61 2 91446895
Email: info@thumbaccess.com
Web: www.thumbaccess.com

Time One Inc
9909 Roosevelt Road, Suite 200, Westchester, IL 60154, USA
Tel: +1 708 216 9890
Fax: +1 708 216 9960
Email: sales@time-one.com
Web: www.time-one.com
Marketing Manager: Kent Talamo

TMA Associates
PO Box 570730, Tarzana, CA 91357-0730, USA
Tel: +1 818 708 0962
Fax: +1 818 345 2980
Email: info@tmaa.com
Web: www.tmaa.com
President: William Meisel

TMS Inc
PO Box 395, 2091 Nooseneck Hill Road, Coventry, RI 02816, USA
Tel: +1 401 397 7550
Fax: +1 401 397 7548
Email: info@tmsbiometrics.com
Web: www.tmsbiometrics.com
President & CEO: Patricia C Gagne

T-NETIX Inc
2155 Chenault Drive, Suite 410, Carrollton, TX 75006, USA
Tel: +1 972 241 1535
Email: information@t-netix.com
Web: www.t-netix.com
President & CEO: Thomas Larkin

Touchless Sensor Technology AG
Filchnerstr 16, 89231 Neu-Ulm, Germany
Tel: +49 731 800 18 0
Fax: +49 731 800 18 18
Email: sales@tst-ag.de
Web: www.tst-ag.de
Contact: Karin Lehner

TSSI
Rutland House, Hargreaves Road, Groundwell Industrial Estate, Swindon, Wiltshire SN25 5AZ, UK
Tel: +44 1793 747740
Fax: +44 1793 747701
Email: robin.horne@tssi.co.uk
Web: www.tssi.co.uk
International Sales Manager: Robin Horne

U

Ultra-Scan Corporation
4240 Ridge Lea Rd, Amherst, New York, NY 14226, USA
Tel: +1 716 832 6269
Email: info@ultra-scan.com
Web: www.ultra-scan.com
CEO: Terry Dunlap

UNIDAS
Germany
Tel: +49 6172 90 24 60
Fax: +49 6172 90 24 70
Email: unidas@aol.com
Web: www.unidas.com

Union Community
Hyundai Topics B/D 5th, 7th, 11th Floor, 44-3, Pangi-dong, Songpa-ku, Seoul 138 – 050, South Korea
Tel: +82 2 6488 3000
Fax: +82 2 6488 3099
Email: sale@unioncomm.co.kr
Web: www.unioncomm.co.kr

Unisys
Township Line & Union Meeting Rds-A' Unisys Way, Blue Bell, PA 19424, USA
Tel: +1 215 986 4011
Web: www.unisys.com
President & General Manager: Greg Baroni

V

Valyd Inc
2099 Gateway Place, Suite 200, San Jose, CA 95110, USA
Tel: +1 408 436 1000
Fax: +1 408 904 7044
Email: sales@Valyd.com
Web: www.valyd.com
CEO: Srinivas Sreeramaneni

Veridicom Inc
1248 Reamwood Ave, Sunnyvale, CA 94089, USA
Tel: +1 408 588 9400
Fax: +1 408 734 0308
Email: info@veridicom.com
Web: www.veridicom.com
President & CEO: Naeem Zafar

Veritel Corporation
35 East Wacker Drive, Suite 690, Chicago, IL 60601, USA
Tel: +1 312 803 5000
Fax: +1 312 803 3311
Email: info@veritelcorp.com
Web: www.veritelcorp.com

VeriTouch Ltd
New York, USA
Tel: +1 917 378 2525
Email: gb@veritouch.com
Web: www.veritouch.com
CEO: Gary E Brant

VeriVoice Inc
5 Vaughn Drive, Princeton, NJ 08540, USA
Tel: +1 609 452 9220
Email: info@verivoice.com
Web: www.verivoice.com

Viisage Technology Inc
30 Porter Rd, Littleton, MA 01460, USA
Tel: +1 978 952 2200
Fax: +1 978 952 2225
Email: info@viisage.com
Web: www.viisage.com
CEO & President: Bernard Bailey

VisionSphere Technologies Inc
Building M-50, National Research Council, 1200 Montreal Road, Ottawa, Ontario, Canada K1A 0R6
Tel: +1 613 740 0245
Fax: +1 613 993 8054
Email: skhan@visionspheretech.com
Web: www.visionspheretech.com
CEO: Sal Khan

Vocalis Group plc
Chaston House, Mill Court, Great Shelford, Cambridge, Cambridgeshire CB2 5LD, UK
Tel: +44 1223 846177
Fax: +44 1223 846178
Email: enquiries@vocalis.com
Web: www.vocalis.com
Head of Marketing: Philippa Buttle

Vocent Solutions Inc
2023 Stierlin Court, Mountain View, CA 94043, USA
Tel: +1 650 938 3663
Fax: +1 650 938 3019
Email: info@vocent.com
Web: www.vocent.com
CEO: Chuck Buffum

Voice Security Systems Inc
24591 Seth Circle, Dana Point, CA 92629, USA
Tel: +1 949 493 4030
Fax: +1 949 388 7714
Email: info@voice-security.com
Web: www.voice-security.com
President & CEO: David Eisenbach

Voicevault
Clonmel House, 17 Harcourt Street, Dublin 2, Ireland
Tel: +353 1 603 9500
Fax: +353 1 478 4478
Email: info@voicevault.com
Web: www.voicevault.com
CEO: Patrick Keaney

W

Watchvision Inc
Dongjin Building 3F, Dongseon-dong 3-ga, Seongbuk-ku, Seoul 136-053, South Korea
Tel: +82 2 928 8783
Fax: +82 2 928 8784
Email: sales@watchvision.com
Web: www.watchvision.com

Wondernet Ltd
Kibbutz Givat Hashlosha 48800, Israel
Tel: +972 3 9036860
Fax: +972 3 9304925
Email: info@wondernet.co.il
Web: www.wondernet.co.il
Contact: Dr Alon Adler

Y

Yasmin Teknologi Sdn Bhd
Pusat Perdagangan Phileo Damansara 1, Block B, Unit 3A03, No 9, Jalan 16/11, Off Jalan Damansara, Petaling Jaya, Selangor 46350, Malaysia
Tel: +603 79577066
Fax: +603 76607677
Email: admin@yasmin.com.my
Web: www.yasmin.com.my

Yuean Biometrics
Yuean Building, No 62-68, Huajing Road, Tianhe District, Guangzhou 510630, China
Tel: +86 20 38761888
Fax: +86 20 38761008
Email: sales@yuean.com.cn
Web: www.yuean.com.cn
Senior Marketing Executive: Philip Tse

Z

Zefyr Technologies
Zone Industrielle des Soeurs, 70 Avenue Bernadotte, Rochefort Sur Mer 17300, France
Tel: +33 5 46 878 454
Fax: +33 5 46 878 455
Email: contact-europe@zefyr.com
Web: www.zefyr.com

ZN Vision Technologie AG
Universitatsstrasse 160, 44801 Bochum, Germany
Tel: +49 234 9787 0
Fax: +49 234 9787 77
Email: info@zn-ag.com
Web: www.zn-ag.com
CEO: Marcel Yon

5.2 Breakdown of Companies by Technology, then by Country

AFIS Systems

Austria
Biometrix International

Germany
Dermalog

Malaya
Yasmin Teknologi

South Africa
FACE Technologies

South Korea
Hyundai Information Technology

Spain
Advanced Software Technologies (AST)

USA
Cogent Systems
East Shore Technologies
Lockheed Martin Information Technology
NEC Technologies
Printrak International
Sagem Morpho
Ultra-Scan

Dynamic Signature Verification

Australia
e-Com

Canada
Cavio
Security Biometrics
Sign@Tech

France
Ingenico

Germany
Softpro
Software Professional

India
ILI Technologies

Israel
Wondernet

Switzerland
App Informatik Davos

UK
Florentis
MotionTouch

USA
BioNetrix System
Communication Intelligence
Cyber-SIGN
SQN Banking Systems
Unisys
Valyd

Facial Recognition

Australia
CSIRO

Canada
3M-AiT
AcSys Biometrics
AND Corporation
ComnetiX Computer Systems
Imagis Technologies
Nexus
VisionSphere Technologies

France
AGMA

Germany
Bergdata
Cognitec Systems
C-Vis
Plettac Electronic Security
UNID
ZN Vision Technologies

Hong Kong
Hectrix
iGuard Security Systems

India
Axis Software

Italy
Biometrika

Norway
Biometri Systemer

Russia
Optical Recognition Objectives

South Africa
Intervid International

South Korea
IDTECK
Sintec
Watchvision

Switzerland
A4Vision

Taiwan
Dialog Communication Systems
Symtron Technology

UK
Aurora Computer Services
Dectel Security
EyeNetWatch.com
IdentAlink
Image Metrics
Integris (now Steria)
Neurodynamics
Neusciences
Ringdale UK
Securicor IS
Serco
TSSI

USA
Aurora Biometrics
Biometrica Systems
BioNetrix System
BioThentica
Business Systems Engineering
Digimarc ID Systems
FaceBase
Geometrix
Graphco Technologies
IBMC
Identix
ImageWare Systems
Imaging Automation
IQ Biometrix
Omron Technology Ventures Group
Oxford Micro Devices
Qvoice
Sense Holdings
Unisys
Viisage Technology

Fingerprint

Argentina
Ex Cle

Australia
Automa
Banque-Tec International
ComSec Enterprises
Fingerprint Technologies
Protocom Development Systems
Rofin Australia
ThumbAccess

Austria
Biometrix International

Belgium
Keyware Technologies

Brazil
Griaule Fingerprint Recognition

Canada
3M-AiT
Affinitex
ALMEX
Bioscrypt
Cansec Systems
ComnetiX Computer Systems
CompuBlox
DelSecur
GEZ Microsystems
Kinetic Sciences
Labcal Technologies
Simple Technology

China
Miaxis Biometrics
Yuean Biometrics

Czech Republic
Digitus

France
Zefyr

Germany
Astro Datensysteme AG
Bergdata
Biometric Partners Europe
BioTrust Project
BrainTrust Technology
Delsy Electronic Component
Fraunhofer Institute for Integrated Circuits
Heimann Biometric Systems
Infineon Technologies
ORGA Kartensysteme
Siemens Biometrics
TeleTrust
Touchless Sensor Technology

Hungary
Guardware

India
Axis Software
BioEnable Technologies
Fusion Tech
ILI Technologies
Jaypeetex Engineering
Print Electronics

Israel
IQS Biometric Solutions
TeKey Research Group

Italy
Biometrika
Digicom
Green Bit
MESA
Neuricam

Japan
Techno Imagia

Jordan
Ideal Software

Lithuania
Neurotechnoligija

Nigeria
Kingshad International

Norway
Biometri Systemer
Idex

Poland
ASTRONTECH
Cassini Advanced Systems
Optel

Portugal
Bioacesso
Proglobo Biometrics Technologies

Puerto Rico
Biometrics Imagineering

Russia
Sonda

Singapore
Autostar Technology

South Africa
Avalanche Technologies
Biometrics.co.za
Cyber-Card

South Korea
Cecrop
FIST

Hunno Technologies
Hyundai Information Technology
IDTECK
Keico Hi Tech
NITGen
Security Information Technology
Suprema
Union Community

Spain
Intuate Biometrics

Sweden
Fingerprint Cards
Precise Biometrics

Switzerland
Bio-Sec
Idencom
IDIAP
PrintScan International

Taiwan
Arithmetic Guarding Equipments (AGE)
Excel Systems
Startek Engineering

The Netherlands
Bell ID

UK
Accu-Tech Systems
Association for Biometrics
Audata
Axxess Identification
Bannerbridge
Bell Security
BTG International
Datastrip
EyeNetWatch.com
Integris (now Steria)
Mason Vactron
Navigator Solutions
Neurodynamics
Photobase
Ringdale UK
Securicor IS
TSSI

USA
Access Controls
ActivCard
Alternative Computer Technology
Applied Biometrics Products
Atmel
AuthenTec
Aware
Biocentric Solutions
BioconX
BioIdentix
BIO-key International
BioLink Technologies International
Biometric Access
Biometric Associates
Biometrics Direct
BioNetrix System
BioPay
BioThentica
Cherry
Clinisync
Control Module
Cross Match Technologies
Daon
Data Management
Datacard Group
Datakey Electronics
DataTreasury
DBA Systems
Digimarc ID Systems
Digital Descriptor Systems
Digital Persona
eCryp
Ethentica
Exact Identification
exResource
Fidelica Microsystems
Financial Information Systems
FingerPrint USA
FingerSec
I/O Software
IBMC
ID Technologies
IDentification Systems
Identix
IDynta Systems
Inception Technologies
Indivos
International Biometrics Group
KeyTronicEMS
Microsoft
National Biometric Test Center
NIST
Novell (NMAS)
Prism Consulting
Privacy Curtain
Qvoice
Rayco Security
SAFLINK
Sagem Morpho
SecuGen
SecureTech Solutions
Sense Holdings

SentryKIDS FingerTIPS
Siemens PSE TecLab
Smart Biometrics
STMicroelectronics
Stromberg
SyntheSys Secure Technologies
TASC
Tennyson Biometric Systems
The International Biometric Society
The Phoenix Group
TMS
Veridicom
VeriTouch

Finger Geometry

Switzerland
BioMet Partners

Taiwan
Excel Systems

UK
Accu-Tech Systems

USA
Accu-Time Systems
Stromberg

Hand Geometry

Canada
ALMEX

Czech Republic
Digitus

Israel
Opticom Technologies

South Korea
Nextern

UK
Control for Doors

USA
Advanced Biometrics
BioLynx
Inception Technologies
Konetix
NOVUS Technology (formerly LARES)
Rayco Security
Recognition Systems
Time One

Iris Recognition

Argentina
Ex Cle

Belgium
Keyware Technologies

Canada
CompuBlox

France
Zefyr

Germany
Fraunhofer Institute for Integrated Circuits
Orga Card Systems
SD Industries

India
Axis Software
Jaypeetex Engineering

Israel
IQS Biometric Solutions
TeKey Research Group

Japan
Oki Electric Industry

Jordan
Ideal Software

Lithuania
Neurotechnoligija

Poland
ASTRONTECH
Cassini Advanced Systems

Portugal
Bioacesso
Proglobo Biometrics Technologies

South Africa
Cyber-Card
Intervid International

South Korea
Senex Technologies

Switzerland
IDIAP

The Netherlands
Bell ID

UK
Aditech
Audata
BTG International
EyeNetWatch.com
Integris (now Steria)
Panasonic UK
Securicor IS

USA
BioconX
Biometrics Direct
Business Systems Engineering
Daon
Datacard Group
DataTreasury
EyeTicket
I/O Software
IBMC
International Biometrics Group
Iridian Technologies
Microsoft
National Biometric Test Center
NIST
Novell (NMAS)
Prism Consulting
Rayco Security
SAFLINK
SecureTech Solutions
TASC
TMS
Unisys
VeriTouch

Speaker Verification

Belgium
ScanSoft BVBA

Canada
AcSys Biometrics
Nexus
Ottawa Telephony Group
Security Biometrics

India
ILI Technologies

Ireland
Voicevault

Israel
Configate
Sentry Com

Italy
Loquendo

Lithuania
Technogama UAB

Russia
Speech Technology Center

Switzerland
Bio-Sec
IDIAP

Taiwan
Dialog Communication Systems

UK
Domain Dynamics
Ensigma Technologies
Securicor IS
Vocalis

USA
Anovea Authentication Technology
Audiopoint
BioNetrix System
Grapho Technologies
I/O Software
Indivos
Interactive Products
International Electronics
Intervoice
J. Markowitz Consultants
Lone Wolf Software
Motorola Semiconductor Products
NCT Group
Nuance Communications
Persay
Qvoice
Sense Holdings
Sonetech
SpeechWorks
TMA Associates
T-NETIX
Unisys
Veritel
VeriVoice
Vocent Solutions
Voice Security Systems

Other Technologies

Belgium
EyeDentify Europe (Retina)

UK
Neusciences (Vein)

USA
AIMS Technology (Nail Bed)
Inroad Solutions (Keystroke Dynamics)
Lumidigm (Spectral Biometrics)
Net Nanny Software International (Keystroke Dynamics)
Retinal Technologies (Retina)

5.3 Breakdown of Companies by Country

Argentina
Ex Cle

Australia
Automa Pty
Banque-Tec International
ComSec Enterprises
CSIRO
e-Com
Fingerprint Technologies
Protocom Development Systems
Rofin Australia
ThumbAccess

Austria
Biometrix International

Belgium
EyeDentify Europe
Keyware Technologies
ScanSoft BVBA

Brazil
Griaule Fingerprint Recognition

Canada
3M-AiT
AcSys Biometrics
Affinitex
ALMEX
AND
Bioscrypt
Cansec Systems
Cavio
ComnetiX Computer Systems
CompuBlox
DelSecur
GEZ Microsystems
Imagis Technologies
Inroad Solutions
Kinetic Sciences
Labcal Technologies
Nexus
Ottawa Telephony Group
Security Biometrics
Sign@Tech
Simple Technology
The Ottawa Technology Group
VisionSphere Technologies

China
Miaxis Biometrics
Yuean Biometrics

Czech Republic
Digitus

France
AGMA
Ingenico
Zefyr

Germany
Astro Datensysteme
Bergdata
Biometric Partners Europe
BioTrust Project
BrainTrust Technology
Cognitec Systems
C-Vis
Delsy Electronic Components
Dermalog
Fraunhofer Institute for Integrated Circuits
Heimann Biometric Systems
Infineon Technologies
Orga Card Systems
Plettac Electronic Security
SD Industries
Siemens Biometrics
Softpro
Software Professional
TeleTrust
Touchless Sensor Technology
UNIDAS
ZN Vision Technologies

Hong Kong
Hectrix
iGuard Security Systems

Hungary
Guardware

India
Axis Software
BioEnable Technologies
Fusion Tech
ILI Technologies
Jaypeetex Engineering
Print Electronics

Ireland
Voicevault

Israel
Configate
IQS Biometric Solutions
Opticom Technologies
Sentry Com
TeKey Research Group
Wondernet

Italy
Biometrika
Digicom
Green Bit
Loquendo
MESA
Neuricam

Japan
Oki Electric Industry
Techno Imagia

Jordan
Ideal Software

Lithuania
Neurotechnoligija
Technogama

Malaya
Yasmin Teknologi

Nigeria
Kingshad International

Norway
Biometri Systemer
Idex

Poland
ASTRONTECH
Cassini Advanced Systems
Optel

Portugal
Bioacesso
Proglobo Biometrics Technologies

Puerto Rico
Biometrics Imagineering

Russia
Optical Recognition Objectives
Sonda
Speech Technology Center

Singapore
Autostar Technology

South Africa
Avalanche Technologies
Biometrics.co.za
Cyber-Card
FACE Technologies
Intervid International

South Korea
Cecrop
FIST
Hunno Technologies
Hyundai Information Technology
IDTECK
Keico Hi Tech
Nextern
NITGen
Security Information Technology
Senex Technologies
Sintec
Suprema
Union Community
Watchvision

Spain
Advanced Software Technologies (AST)
Intuate Biometrics

Sweden
Fingerprint Cards
Precise Biometrics

Switzerland
A4Vision
App Informatik Davos
BioMet Partners
Bio-Sec

Idencom
IDIAP
PrintScan International

Taiwan
Arithmetic Guarding Equipments (AGE)
Dialog Communication Systems
Excel Systems
Startek Engineering
Symtron Technology

The Netherlands
Bell ID

UK
Accu-Tech Systems
Aditech
Association for Biometrics
Audata
Aurora Computer Services
Axxess Identification
Bannerbridge
Bell Security
Biometric Technology Today
BTG International
Control for Doors
Datastrip
Dectel Security
Domain Dynamics
Ensigma Technologies
EyeNetWatch.com
Florentis
IdentAlink
Image Metrics
Integris (now Steria)
Mason Vactron
MotionTouch
Navigator Solutions
Neurodynamics
Neusciences
Panasonic UK
Photobase
Ringdale UK
Securicor IS
Serco
TSSI
Vocalis

USA
Access Controls
Accu-Time Systems
ActivCard
Advanced Biometrics
AIMS Technology
Alternative Computer Technology
Anovea Authentication Technology
Applied Biometrics Products
Atmel
Audiopoint
Aurora Biometrics
AuthenTec
Aware
Biocentric Solutions
BioconX
BioIdentix
BIO-key International
BioLink Technologies International
BioLynx
Biometric Access
Biometric Associates
Biometric Consortium
Biometric Digest
Biometrica Systems
Biometrics Direct
BioNetrix System
BioPay
BioThentica
Business Systems Engineering
Cherry
Clinisync
Cogent Systems
Communication Intelligence
Control Module
Cross Match Technologies
Cyber-SIGN
Daon
Data Management
Datacard Group
Datakey Electronics
DataTreasury
DBA Systems
Digimarc ID Systems
Digital Descriptor Systems
Digital Persona
East Shore Technologies
eCryp
Ethentica
Evive
Exact Identification
exResource
EyeTicket
FaceBase
Fidelica Microsystems
Financial Information Systems
FingerPrint USA
FingerSec
Geometrix
Grapho Technologies
I/O Software
IBIA
IBMC
ID Technologies

IDentification Systems
Identix
IDynta Systems
ImageWare Systems
Imaging Automation
Inception Technologies
Indivos
Interactive Products
International Biometrics Group
International Electronics
InterVoice-Brite
IQ Biometrix
Iridian Technologies
J. Markowitz Consultants
KeyTronicEMS
Konetix
Lockheed Martin Information Technology
Lone Wolf Software
Lumidigm
Microsoft
Motorola Semiconductor Products
National Biometric Test Center
NCT Group
NEC Technologies
Net Nanny Software International
NIST
Novell (NMAS)
NOVUS Technology (formerly LARES)
Nuance Communications
Omron Technology Ventures Group
Oxford Micro Devices (OMDI)
Persay
Polaroid ID
Printrak International
Prism Consulting
Privacy Curtain
Qvoice
Rayco Security
Recognition Systems
Retinal Technologies
SAFLINK
Sagem Morpho
SecuGen
SecureTech Solutions
Sense Holdings
SentryKIDS FingerTIPS
Siemens PSE TecLab
Smart Biometrics
Sonetech
SpeechWorks
SQN Banking Systems
STMicroelectronics
Stromberg
SyntheSys Secure Technologies
TASC
Tennyson Biometric Systems
The International Biometric Society
The Phoenix Group
Time One
TMA Associates
TMS
T-NETIX
Ultra-Scan
Unisys
Valyd
Veridicom
Veritel
VeriTouch
VeriVoice
Viisage Technology
Vocent Solutions
Voice Security Systems

5.4 Breakdown of Companies by Geographic Region

Africa
Avalanche Technologies
Biometrics.co.za
Cyber-Card
FACE Technologies
Intervid International
Kingshad International

Asia
Arithmetic Guarding Equipments
Automa Pty
Autostar Technology
Axis Software
Banque-Tec International
BioEnable Technologies
Cecrop
ComSec Enterprises
CSIRO
Dialog Communication Systems
e-Com
Excel Systems
Fingerprint Technologies
FIST
Fusion Tech
Hectrix
Hunno Technologies

Hyundai Information Technology
IDTECK
iGuard Security Systems
ILI Technologies
Jaypeetex Engineering
Keico Hi Tech Inc
Miaxis Biometrics
Nextern
NITGen
Oki Electric Industry
Optical Recognition Objectives
Print Electronics
Protocom Development Systems
Rofin Australia
Security Information Technology
Senex Technologies
Sintec Corporation
Sonda
Speech Technology Center
Startek Engineering
Suprema
Symtron Technology
Techno Imagia
ThumbAccess
Union Community
Watchvision
Yasmin Teknologi
Yuean Biometrics

Europe
A4Vision
Accu-Tech Systems
Aditech Limited
Advanced Software Technologies
AGMA
App Informatik Davos
Association for Biometrics
Astro Datensysteme
ASTRONTECH
Audata
Aurora Computer Services
Axxess Identification
Bannerbridge
Bell ID
Bell Security
Bergdata
Bioacesso
BioMet Partners
Biometri Systemer
Biometric Partners Europe
Biometric Technology Today
Biometrika
Biometrix International
Bio-Sec
BioTrust Project
BrainTrust Technology
BTG International
Cassini Advanced Systems
Cognitec Systems
Control for Doors
C-Vis
Datastrip
Dectel Security
Delsy Electronic Components
Dermalog
Digicom
Digitus
Domain Dynamics
Ensigma Technologies
EyeDentify Europe
EyeNetWatch.com
Fingerprint Cards
Florentis
Fraunhofer Institute for Integrated Circuits
Green Bit
Guardware
Heimann Biometric Systems
Idencom
IdentAlink
Idex
IDIAP
Image Metrics
Infineon Technologies
Ingenico
Integris (now Steria)
Intuate Biometrics
Keyware Technologies
Loquendo
Mason Vactron
MESA
MotionTouch
Navigator Solutions
Neuricam
Neurodynamics
Neurotechnoligija
Neusciences
Optel
Orga Card Systems
Panasonic UK
Photobase
Plettac Electronic Security
Precise Biometrics
PrintScan International
Proglobo Biometrics Technologies
Ringdale UK
ScanSoft BVBA
SD Industries
Securicor IS
Serco
Siemens Biometrics
Softpro
Software Professional

Technogama UAB
TeleTrust
Touchless Sensor Technology
TSSI
UNIDAS
Vocalis
Voicevault
Zefyr
ZN Vision Technologies

Middle East
Configate
Ideal Software
IQS Biometric Solutions
Opticom Technologies
Sentry Com
TeKey Research Group
Wondernet

North & South America
3M-AiT
Access Controls
Accu-Time Systems
AcSys Biometrics
ActivCard
Advanced Biometrics
Affinitex
AIMS Technology
ALMEX
Alternative Computer Technology
AND
Anovea Authentication Technology
Applied Biometrics Products
Atmel
Audiopoint
Aurora Biometrics
AuthenTec
Aware
Biocentric Solutions
BioconX
BioIdentix
BIO-key International
BioLink Technologies International
BioLynx
Biometric Access
Biometric Associates
Biometric Consortium
Biometric Digest
Biometrica Systems
Biometrics Direct
Biometrics Imagineering
BioNetrix System
BioPay
Bioscrypt
BioThentica
Business Systems Engineering
Cansec Systems
Cavio
Cherry
Clinisync
Cogent Systems
Communication Intelligence
CommetiX Computer Systems
CompuBlox
Control Module
Cross Match Technologies
Cyber-SIGN
Daon
Data Management
Datacard Group
Datakey Electronics
DataTreasury
DBA Systems
DelSecur
Digimarc ID Systems
Digital Descriptor Systems
Digital Persona
East Shore Technologies
ECryp
Ethentica
Evive
Ex Cle
Exact Identification
exResource
EyeTicket
FaceBase
Fidelica Microsystems
Financial Information Systems
FingerPrint USA
FingerSec
Geometrix
GEZ Microsystems
Grapho Technologies
Griaule Fingerprint Recognition
I/O Software
IBIA
IBMC
ID Technologies
IDentification Systems
Identix
IDynta Systems
ImageWare Systems
Imaging Automation
Imagis Technologies
Inception Technologies
Indivos
Inroad Solutions
Interactive Products
International Biometrics Group
International Electronics
InterVoice-Brite
IQ Biometrix

Iridian Technologies
J. Markowitz Consultants
KeyTronicEMS
Kinetic Sciences
Konetix
Labcal Technologies
Lockheed Martin Information Technology
Lone Wolf Software
Lumidigm
Microsoft
Motorola Semiconductor Products
National Biometric Test Center
NCT Group
NEC Technologies
Net Nanny Software International
Nexus
NIST
Novell (NMAS)
NOVUS Technology
Nuance Communications
Omron Technology Ventures Group
Ottawa Telephony Group
Oxford Micro Devices (OMDI)
Persay
Polaroid ID
Printrak International
Prism Consulting
Privacy Curtain
Qvoice
Rayco Security
Recognition Systems
Retinal Technologies
SAFLINK
Sagem Morpho
SecuGen
SecureTech Solutions
Security Biometrics
Sense Holdings
SentryKIDS FingerTIPS
Siemens PSE TecLab
Sign@Tech
Simple Technology
Smart Biometrics
Sonetech
SpeechWorks
SQN Banking Systems
STMicroelectronics
Stromberg
SyntheSys Secure Technologies
TASC
Tennyson Biometric Systems
The International Biometric Society
The Ottawa Technology Group
The Phoenix Group
Time One
TMA Associates
TMS
T-NETIX
Ultra-Scan
Unisys
Valyd
Veridicom
Veritel
VeriTouch
VeriVoice
Viisage Technology
VisionSphere Technologies
Vocent Solutions
Voice Security Systems

6 Appendix

6.1 Understanding Biometric Technology

6.1.1 Introduction

Biometrics are measurable physiological or behavioural characteristics that can be used to verify the identity of an individual. There are a wide variety of different types of biometric available for measurement. The most popular include a person's fingerprint, face, hand, iris, voice, signature, retina or typing rhythm. However other biometrics are under research, such as gait recognition (the way a person walks), earprint analysis, DNA recognition or even recognising a person from their spectral characteristics (which is a measure of the composition and structure of a person's skin and underlying tissue).

The biometric industry produces systems that use a person's biometric, or a combination of biometrics, to automatically check that person's identity.

6.1.2 Operation of Biometric Systems

6.1.2.1 Enrolment

Before an individual can be verified using a biometric system, they must first go through an enrolment procedure. Typically, enrolment involves an individual presenting a sample of the biometric to be used by the system – their signature, fingerprint, voice or other biometric. This sample is then used to generate a template.

The template can either be stored within a verification terminal, if it is a stand-alone system, within a central database for a networked or on-line system, or on a plastic card or document, which the enrolled individual carries with them.

Once the template has been created, a way of linking the individual to their template must be found. This can be done in a number of ways, for example, by issuing the individual with a PIN or giving them a card with a unique serial number. Either method enables an index to be created which can link the individual to their template.

6.1.2.2 Verification Versus Identification

When the individual next needs to use the biometric system – for example to gain access to a building or to make a purchase at a biometrically secured point of sale – they simply enter their PIN or present their card and give a new sample of their individual characteristic. Their biometric template is called up and compared with the new sample. Provided the template and the new sample are sufficiently similar to each other, the system will verify them and allow them to make a transaction or access the protected area.

It is this use of a template identifier that differentiates verification (or authentication) – the process in which biometric devices are most commonly used – from identification. In an identification system, there is no template identifier. The system takes the new sample and compares it against all of the samples on file. The sample, or samples, that most closely match are pulled out of the memory and given as the most likely people to whom the sample belongs.

This process is the way in which AFIS (Automated Fingerprint Identification Systems) systems are operated. AFIS systems are most frequently used to automate the identification of possible culprits for crimes where a fingerprint has been left behind. However they are increasingly being used in large welfare systems to check whether a person is cheating by making multiple claims while using assumed identities. Employee background checks are also becoming popular using AFIS systems.

Identification requires far more processing power than verification, where a sample is compared against only one other record.

6.1.2.3 No 100% Match

A major difference between biometric verification and PINs or ID numbers is that a biometric system does not give an absolute 'yes' or 'no' answer when asked to verify an individual. Where a PIN is either '1234' or it is not, a biometric characteristic will vary from reading to reading.

Although most people's signature will always remain basically the same, even the most consistent individual will slightly vary their signature every time it is written. With biometric systems using optical imaging techniques to capture a feature such as a fingerprint, the quality of the captured image will vary each time a reading is taken. So although the fingerprint or retinal pattern itself does not vary between readings, the resultant image will not be an exact duplicate of that stored on file.

A biometric system, therefore, cannot verify an individual's identity with 100% certainty. Instead the system will give a probability level that this is the person they claim to be. Of course, this probability can run very close to one.

All biometric systems must, therefore, incorporate a degree of tolerance. If the system were to require an almost 100% match between the template and the new reading, then the vast majority of legitimate users would be rejected. In practice biometric systems require only that the new reading and the template are a close match.

By lowering the tolerance levels, however, another problem occurs – the tolerance might be low enough to allow entry to a small number of impostors. Biometric systems can, therefore, be set to different tolerances according to requirements.

For instance, in a banking application, the system may be set so that a relatively high percentage of impostors are accepted but very few legitimate customers are rejected, thus ensuring that the bank does not have scores of dissatisfied customers. For a high security application, however, such as protecting access to a military base, the same system could be set to allow virtually no impostors. This would result in some legitimate personnel being refused entry and they could face the inconvenience of having to try two or three times to gain access. (As long as they do finally get access!)

6.1.2.4 False Accept and False Reject Rates

There must, therefore, be a trade-off between allowing fraudsters into the biometric system and refusing entry to legitimate users – you cannot have it both ways. Whether it is more important to never let in an impostor or to never refuse a legitimate user can determine the type of biometric technology chosen for an application, as well as the particular product.

Some biometric approaches are inherently more accurate when it comes to rejecting fraudsters than others, but are more difficult for legitimate users to be verified – they have a very low False Acceptance Rate (FAR), which is also known as a Type 2 Error Rate. The FAR represents the proportion of impostors allowed into a system.

Others are better at correctly verifying legitimate users but let in more fraudsters – they have a low False Rejection Rate (FRR), also known as a Type 1 Error Rate. FRR represents the proportion of people falsely rejected by the system.

The point where the FRR numerically equals the FAR is known as the Equal Error Rate. This normally represents a more realistic measure of performance than either of the individual rates.

6.1.2.5 Cost

The accuracy of a biometric device, while obviously important, is by no means the only factor that affects the choice of a biometric system. Of the others, the cost of implementing the system is a factor of major importance. Prices have been steadily falling in most sectors of the biometrics market over the last decade. Where the cost was initially prohibitive to all but the highest security applications, biometrics are now expanding into more medium and low security applications as the price drops.

The industry is now on the verge of breaking through the price barrier and launching itself into the mass market, especially in the area of computer security, where biometric devices can be purchased for less than US$50.

6.1.2.6 Public Acceptance

Of equal importance in a biometric consumer application is the acceptability of the technology to those who will have to use it. This is true of both work and consumer environments.

If a loyal customer of one retailer or bank does not wish to use the biometric system that has been implemented, then they can simply take their money elsewhere. In general, the more transparent the technology is to the user, then the more likely it is to be accepted.

In environments where biometrics are used for surveillance then public acceptance becomes crucial, as the system works on everybody within its range. This type of system has the potential to create significant public backlash, as it did in the USA, where its use was used at sporting events and is now deployed on the streets of certain downtown areas.

6.1.2.7 Other Factors

There are a number of other factors that also need to be considered. The memory capacity needed to store a biometric template varies greatly from product to product and technology to technology. This is an important consideration in mass-market applications, where it may be desirable to store the template on a plastic card.

The time taken for the verification process to be performed can also be important. It could make the difference between queues at the supermarket checkout or bottlenecks at the main entrance to an office building.

The size of the verification device and its physical reliability once implemented also need to be taken into account. Biometric devices are getting smaller, with chip-sized scanners now available for installation in mobile telephones and computer keyboards.

6.1.3 Template Storage Options

An important consideration for a biometric system developer to decide is the location of the enrolees' templates. Should the template be stored in a large central database (worrying to some people) or on a card belonging to the user?

Generally storage issues depend on the type of application, on the biometric device being used and on the number of people to be enrolled. While storing templates on plastic cards for verification at the point of sale for banking transactions might make economic sense, a company would probably not issue plastic cards to 20 employees using a speaker verification system to open the office door.

6.1.3.1 In the Memory of the Biometric Device

Many verification devices have their own memory in which templates can be stored. The number of templates that can be stored within the device itself varies with each supplier's equipment.

When the templates are stored on a stand-alone biometric device, it is much harder for a hacker to gain access to the templates since they do not need to leave the device or get transmitted to another location. With this option no additional costs for telecommunications, computing or plastic cards are incurred either.

6.1.3.2 Central Computer

Storing templates on a biometric device may not be the best solution if users need to be verified at different locations, for example at different offices within a city, at bank branches across a country or at airports across the world.

In these cases, there are two solutions. One is to store the templates remotely, in a central database; the other is to store the templates on tokens (usually cards) carried by the enrolees.

The number of templates that can be stored on a central database is only limited by the capacity of the computer being used. When a user enters their PIN, the computer will locate their template and send it to the biometric device to perform a comparison with the live sample.

Templates held remotely on a central computer can also be downloaded on a daily, weekly or 'as needed' basis to different verification devices. Extra security, such as time and date stamping of the templates as they are transmitted, also needs to be considered. A hacker gaining access to enrolees' templates could seriously compromise the security of the system. An increasingly common security measure is to encrypt templates, while they are in transit and even while they reside on the system.

One factor that has to be considered before opting for remote access to, or long-distance distribution of, templates is whether the telecommunications network is up to the job. Where a telecommunications infrastructure is under-developed, it is not sensible to load telephone lines with massive traffic in biometric templates.

Another consideration is the potential negative reaction of users to a centralised database. The perception that 'big brother' could access and use the templates in some way still exists and could turn people off using the system. Data protection laws must also be consulted to check the system will be legal.

6.1.3.3 Plastic Card or Token

In many systems the decision may be made against storing templates in a central system. In this case, the answer may be to store the template on an identification (ID) device which enrolees can carry around with them.

In an immigration application, for example, biometric data could be printed as a special barcode on passports. Alternatively, a plastic card may be used. The type best suited to a biometric system will depend very much on the actual application and on the amount and type of extra information that has to be stored.

Different types of card are easier to hack and counterfeit than others – all have different prices and some are better suited to certain applications than others.

The most common of plastic cards are magnetic stripe cards. Introduced in the 1950s, billions have been issued, with the majority of people in the developed world having at least one in their wallet. Terminals are already located in many places in the high street, in petrol stations, in public telephones and vending machines.

The main advantage of magnetic stripe cards are their low cost. The downside is that the cards are relatively simple to counterfeit and the magnetic stripe can be easily scratched; it can also be corrupted if it is exposed to a magnetic field.

Smart cards incorporating microprocessors offer the highest level of security to card issuers and cardholders, since they incorporate a 'secret zone' that cannot generally be accessed by anyone or anything except the card's microprocessor. Because of this, information need not be transmitted outside the card and cannot, therefore, be hacked by an outsider. When a smart card is used, the only element that could be accessed is the new sample when it is being transmitted between the biometric device and the smart card. This is made harder where the smart card reader and the biometric scanner are integrated in the same terminal.

Even if a new sample can be accessed during the verification process, the hacker would not know which parts of it were identical with those on the original template and which had changed since the last verification attempt. In fact, a smart card microprocessor can be programmed to reject a sample that shows a 100% match with the template – or indeed a sample identical to one which had been submitted in the past. By the very nature of biometric systems, a 100% similarity between any two samples can only be a forgery.

The biometric scanner itself can even be incorporated into the same device as the processor ie. alongside the cardholder's template. This approach has certainly taken off in the last couple of years with chip-sized scanners and biometrically-enabled smart cards being developed.

There are numerous types of cards that can be used to store templates. These include:

- **Bar code cards** – cheap and capable of carrying around 2-3 Kbytes amounts of information.
- **Magnetic stripe cards** – cheap, common and an existing infrastructure in place.
- **Enhanced magnetic stripe cards** – this encompasses a range of products which incorporate some form of additional anti-counterfeiting technique, such as incorporating special magnetic properties or holographic images.
- **RF-ID cards and tags** – RF-ID cards do not need to be placed within a card reader to be scanned as they contain a tiny radio transmitter that is activated only when it receives a signal of a specific frequency. The communication range varies between a few centimetres and up to several hundred feet.

- **Optical memory cards** – these cards are the same size as credit cards but have one side covered in a reflective, shiny substance. Their main advantage is that they have a huge memory capacity over 4 Mbytes.
- **PCMCIA cards** – These cards can be slotted into a computer in order to either enhance its memory or performance, or to enable it to perform a special function. The cards are the same shape as credit cards but are considerably thicker and have a connection socket along one end. PCMCIA cards can now incorporate a smart card microprocessor and incorporate biometric templates.
- **Smart cards** – Most commonly these comprise of a chip embedded into a credit card-sized piece of plastic. There are typically three types: memory-only chips, logic memory chips and microprocessor-equipped chips. Of these microprocessor-equipped cards are the most sophisticated of all and can be used to perform on-card matching of biometric templates. The typical cost of such a card is US$2–4.

6.1.4 Testing

Proving the performance of a biometric system is one of the big debates within the industry at this point in its evolution. Most manufacturers quote false acceptance and false rejection rates for their different systems, but the conditions that are experienced in laboratories compared with 'real world' environments can be worlds apart.

Factors such as the type of people used to perform the test (habituated users or novices, old or young etc), number of verification attempts allowed and the number of people in a test all affect the end result. Systems proving themselves in real world installations are often seen as more important than error rates.

Testing is widely acknowledged to be a good thing. Vendors need testing to justify products, and to bring them to market. Buyers want testing because they want guarantees, fewer worries and less time spent in evaluation.

6.1.5 Standards

If you were to buy a camera in the USA and then take it to Australia, you would expect Australian camera film to fit into your camera. In the case of camera film, wherever you go, it will fit thanks to international standards governing the dimensions of each roll. Similarly, buyers of biometric systems expect a level of standardisation that will allow them to pick and choose hardware, without having to change their entire system.

Industry standards can help in defining the criteria that manufacturers should work with when designing their system. A developed set of standards demonstrates the increasing maturity of an industry. In the biometric industry, major progress has been made, but there is still work to do.

Pushing biometrics one step closer to plug and play are Applications Programming Interfaces (APIs) which sprung up at the end of 1997. The highest profile API at present is the BioAPI standard.

BioAPI has the backing of a large proportion of industry players and is widely seen as the universal standard that the industry has been waiting for. Indeed it has now been given the status of a US national standard and is moving swiftly forward to become an ISO (International Standards Organisation) standard.

However, along with many other standards for specific applications, perhaps the most important to watch is an alternative standard to be launched by software giant Microsoft for use with its Windows operating system in the near future. So the final standards landscape is still unclear.

A full update on standards is given in Appendix 6.3.

6.1.6 Biometric Technologies

The following section will review the basics of the main biometric technologies including:

- Fingerprint
- Eye
- Hand
- Face
- Voice
- Signature

6.1.6.1 Fingerprint Verification

Although the history of fingerprinting goes back to ancient times, the beginnings of modern day fingerprint comparison techniques can be traced back to 1684. In that year, Nehemiah Grew of the Royal Society in London first noticed that people's fingerprints could be classified by a number of key patterns that include loops, whorls and arches.

Today, as well as being classified by the main characteristics, fingerprints are also classified by smaller characteristics, known as minutiae, which occur at the terminations and branches of ridges on the fingerprint area. One of the first users of fingerprints for identification was Juan Vecutich, an Argentinean police officer who set up a national fingerprint system in 1896. Soon afterwards in 1901, Sir Edward Henry established a fingerprint branch at the UK's Scotland Yard.

Fingerprint systems are now the most widely used biometric technology in the world. The main users of fingerprints are the world's police forces who started fingerprinting offenders in the mid-1880s. If a fingerprint is to be used as evidence in a court of law, a certain number of characteristics of the print found at the scene of a crime must be the same as those found on the print of the suspect. The actual number varies between countries.

Prints of suspects have traditionally been collected using special ink and a card divided into ten sections, one for the print of each finger. Rather than using ink, however, systems are now available which can record a person's fingerprints electronically in essentially the same way that biometric verification systems do –

these are known as 'live-scan' systems – and the fingerprint images can then be printed out directly onto a standard ten print card as required.

As explained earlier, AFIS systems record all the characteristics of a person's fingerprints and then check them against all the other fingerprint records on file in a process known as *identification*. The other main type of fingerprint system takes only the key points of a person's fingerprint to produce a non-reversible template. This is checked against a specific record to verify the person is who they claim to be – the process of *authentication* or *verification*.

6.1.6.1.1 Operation of an Automatic Fingerprint System

A fingerprint pattern covers the area from one side of the fingernail right around the pad of the finger to the other side of the fingernail. Within this area is found a wide variety of distinguishing marks from the major characteristics of arches, loops, whorls and compounds to the minor markings known as minutiae or ridge characteristics. It is these minutiae which are used to make an identification in many automated systems.

An automatic identification system will not normally store an actual picture of the fingerprint, but may calculate and place in memory the co-ordinates of the minutiae found. Because dirt and grease can obscure the minutiae, an enhanced optical process is generally used to read the fingerprint. This approach allows the fingerprint to be read without the finger actually coming into contact with the sensors.

There are also other types of fingerprint sensor that do not use optical methods to capture the fingerprint image, such as capacitive or thermal sensors (based on silicon) or ultrasound devices. With a silicon sensor, an individual's finger acts as one of the plates of a capacitor. The other plate, on the surface of the sensor, consists of a silicon-based chip. This information is converted to a video signal.

With all types of sensor, the required fingerprint characteristics are converted into digital form and stored. When users present themselves for verification, their fingerprint image is again digitised and compared with the digital record.

6.1.6.1.2 Requirements of a Fingerprint Verification System

Public acceptability
Because of the association of fingerprinting with law enforcement, there have been worries, in the past, about the public acceptability of fingerprint verification in the consumer marketplace. But these attitudes now seem to be changing, especially as more systems are installed and the public becomes less fearful of new technologies.

Image quality
While automatically verifying a fingerprint on a clean finger with well defined characteristics and minutiae is a comparatively straightforward process, a system for a mass market application needs to provide flexibility of use and be able to handle a wide variety of print qualities.

The quality of a fingerprint varies according to race, sex, occupation and age. Generally speaking, Caucasians have the easiest prints to read, with the prints becoming less defined in other races. Women tend to have finer print details than men. White collar workers tend to have better quality prints than manual workers, coal miners usually being taken as the worst example.

Age
The ageing process also affects fingerprint verification systems by hardening the skin of the fingertips. Young people have softer skin, in general, and will be required to put more pressure on the fingerprint reader. A reliable system must therefore be able to handle a wide variety of pressures.

Ease of use
Most people find it easiest to use a forefinger but a choice should be made available since a significant number of people have a missing finger or a finger which is too scarred to be usable for verification. The biggest problem seems to lie with the left forefinger.

Circumvention
There are many potential ways that can be used to compromise a fingerprint system – the use of a severed finger is perhaps the most unsavoury, but more often a fake 'gummy' fingerprint can be manufactured. This issue is the subject of a lively debate within the industry at present and many manufacturers are introducing or improving existing 'Live and Well' detection capability which can sense characteristics, such as the body heat of a living finger and reject a false finger. Data being transmitted between the fingerprint terminal and template database are also a vulnerability, but can also be protected as long as sensible safeguards are put in place.

Performance
As with other biometric types, the performance of specific fingerprint verification devices is important and varies between suppliers. Generally speaking fingerprint devices offer a highly accurate method for identifying individuals. However, the ability for fingerprint systems to acquire a reasonable fingerprint image is very important and does vary between devices. Unattended use can also create performance problems as there is a tendency for dirt and grease to accumulate on the sensor surface, creating fingerprint acquisition problems. That said some designs such as ultrasonic or swipe-based sensors (where a person sweeps their fingerprint over a narrow sensor) can bypass this problem. A breed of sensors that look at the 'live' layer under the fingerprint surface also alleviate this problem.

With silicon-based sensors there is also a possible problem of robustness – with factors such as electrostatic discharge or specific environmental conditions potentially causing performance problems. This situation has improved dramatically in the last two years however.

6.1.6.2 The Eye

Two parts of the eye can be measured by separate biometric systems; the iris and the retina. The patterns within the eye – whether they are the retinal vascular pattern or the iris – are as unique to each person as a fingerprint, if not more. Even a person's own two eyes are distinct from each other. As well as the distinctness of these patterns, they also remain stable over a person's life, only being affected by a few diseases. Because of their close links to the brain, the eyes are one of the first parts of the body to decay after death.

The idea that the blood vessels of the retina could be used to identify individuals was published in the New York State Journal of Medicine as long ago as September 1935 by Drs Carleton Simon and Isidore Goldstein. The uniqueness of the retinal pattern was supported in the 1950s by Dr Paul Towers in his investigations of identical twins. He claimed that the retinal vein pattern provided the most difference between twins.

The first retinal scanning units were launched in 1985 by EyeDentify, which quickly established a monopoly in this market. Most of the early units were sold to military and other establishments requiring the high security that these systems offer. The company claimed that its systems have never let in an impostor. Despite this monopoly EyeDentify seems to have hit troubled times. Compounding its problems have been the emergence of a few other companies promising much smaller, cheaper and more user-friendly retina-based systems (although these have yet to hit the market).

Irises are also distinct physical characteristics. Two American ophthalmologists, Dr Leonard Flom and Dr Aran Safir studied the properties of the iris noticing that they have a highly detailed and unique texture and remain stable over periods of many years. The iris pattern itself is made up of furrows, crypts, corona, filaments, pits, freckles, striations and rings.

The first iris scanning systems were launched commercially in 1995 by US company IriScan (now know as Iridian Technologies), which claims to own all rights to the technology.

6.1.6.2.1 Mode of Operation

With the EyeDentification system the user places one eye about two inches away from the device, then stares at a green dot target to bring the eye into focus. Light from a 0.07W source (less than a refrigerator bulb) is reflected off the retina, collected by a rotating lens, before conversion to an analogue signal. The lens scans the retina until it creates a good template.

The newer retina-based systems are being designed to scan a person's retina at arm's length.

Iridian's system uses a monochrome camera with standard video optics to capture the iris image. Early models required close eye positioning, but this has now improved. The iris pattern has 266 independent variables. It is captured and digitised into an IrisCode for storage as a 256 byte template. Smaller devices designed as peripherals for use with PCs have also been launched.

6.1.6.2.2 Requirements of an Eye Scanning System

Public acceptability
Although there were early fears by some end users that retinal scanning would injure their eyes, their worries do not seem to have been passed onto the iris scanning systems. In fact, early comments that iris scanning would be unlikely to take off have been proved to be untrue with the emergence of systems in banks, airports and PCs.

Work and environmental factors
The eye is a very stable part of the human body and is only affected by a few rare diseases. As with all biometrics, however, there are people that the technology will find difficult to accommodate.

Ease of use
Both systems require a degree of user co-operation, although the iris system is less demanding of the user and recent systems developed actually locate and verify the iris pattern automatically.

Circumvention
Because of its close link to the brain, the eye is one of the first parts of the body to decay after death. The vein pattern changes within just a few minutes. This makes it very secure against attack, however low-spec iris systems have been fooled via a photo of the iris – although Iridian is working to correct this phenomenon.

Performance
Eye scanning biometric verification devices are the most accurate of all systems and can work in identification mode on large databases.

6.1.6.3 Hand

Hand measuring technologies divide into three types: the application of fingerprint identification techniques to the palm, the pattern of veins in the hand and the geometrical analysis of part of the hand.

One of the earliest of all biometric devices measured the lengths of the fingers. Called IdentiMat, the device was launched in the early 1970s by Identimation. Production ceased in 1987. Georgia University, in the USA, became the first organisation to upgrade its biometric system when it upgraded from its IdentiMat system to ID3D HandKeys from Recognition Systems in October 1990.

Factors such as dirt and grease tend not to cause a problem for systems which are looking at the geometry of the hand or fingers but may cause problems for those looking at the finer details such as palm prints. Hand and finger geometry units are simple to use with some devices incorporating guide posts to ensure that the hand is correctly placed. Some systems will allow the user to use either hand, whilst others are easier to use with one specific hand.

The recognised leaders in the hand geometry market are Recognition Systems and Biomet Partners. Recognition Systems' device studies the entire hand while Biomet Partners looks at three-dimensional measurements of two fingers from either the left or right hand.

The most successful use for hand verification has been physical access control where systems are being used in a number of prisons. Suppliers have also built a niche in the time and attendance marketplace. Both the Recognition Systems and Biomet Partners systems have been used in high profile public locations, such as the 1996 Olympic Games and the continued enrolment of visitors to Disney Theme Parks in the USA.

Vein pattern analysis stems from patents held by the British Technology Group.

6.1.6.3.1 Requirements of a Hand Geometry System

The performance of independently tested hand geometry verifiers is excellent and the units are not difficult to use. With no precedent for the use of the hand in identifying people, they have the advantage over fingerprint systems of not being adversely prejudged.

Public acceptability
There are few problems with acceptability with this biometric.

Work and environmental factors
Because these devices rely on photographing a silhouette of the hand shape, factors such as dirt and grease do not pose a serious problem. Dirt may, however, obscure some of the pattern on systems that look at the finer details such as creases on the palm.

Ease of use
Although some systems will allow an enrolee to place either hand in the system, this is not the case with all of them. Generally speaking however most of these systems are very easy to use.

Circumvention
Sophisticated models of the bone structure of an authorised hand have some potential for fooling these systems.

Performance
Hand geometry units have the smallest templates of all – only nine bytes is required for one of the systems on the market. One major problem with this biometric type is that it measures a characteristic which is below the skin; the bones. The skin covering the bones of the hand is elastic and, therefore, not constant in its dimensions. This means that a margin for error due to the measurements of the skin must be built in. Problems also arise from people with swollen fingers, perhaps due to arthritis, and also those who occasionally wear large rings. Although a very convenient technology, it is not as good at distinguishing between people as a fingerprint or eye-based system.

6.1.6.4 Facial Recognition

Facial recognition systems are generally based on a database of 'mugshot' photos (typically stored in digital form), that are taken when a person is enrolled. These mugshots – full frontal portraits taken against an uncluttered background – are then available for comparison against later photographic images, taken by a still or video camera.

Although improvements in technology are frequent, most facial recognition systems still suffer from one basic problem; mugshots present the subject in an artificial pose. In matching a 'real life' image against an image held on the database, a facial recognition system has to overcome several difficulties. First, it has to locate the head and face against a cluttered background. Then it has to deal with the fact that the later image will present the face at a different angle and in a different light from when the subject was originally enrolled.

To recognise a face presented on a real life image, the system must first find the face within the area scanned by a still or video camera. Once the face has been found, specific features may have to be located. Systems that compare the face's features, rather than the complete image of the face, usually look for the positioning of the eyes, nose and mouth. In poor light conditions, individual features may not be easily discernible. The edge of a shirt collar, for instance, can merge into the subject's chin. Beards and glasses can cause additional difficulties.

6.1.6.4.1 Operation of a Facial Recognition System

The majority of facial recognition systems on the market today require the user to stand a specific distance away from the system and look straight at a camera. This ensures that the captured image of the face is within a specific size tolerance and keeps the features, such as eyes, in as similar a position each time as possible.

Some systems do not require such rigid placement, and can recognise a person walking through a shopping mall – making it suitable for surveillance applications. Because a number of systems first have to find the outline of the face within the actual captured image, a better result is achieved if the user is not standing in front of a cluttered background. The level of lighting can also affect this process by making shadows and highlights.

Once the image has been captured, it is analysed by the system and a template made. Some systems store an actual image of the face whilst others store code relating to the position and shape of specific features. Better results can be achieved if the operator is able to tell the system exactly where the eyes, or other features, are positioned.

6.1.6.4.2 Requirements of a Facial Recognition System

Public acceptability
As people are already used to being identified by their face, there are no foreseen problems with the use of facial recognition systems for applications such as access control. The one area that does cause some concern is the use of facial recognition with CCTV cameras, thus emulating the 'big brother' scenario. In order to get around potential privacy problems, suppliers have developed systems that mask the faces of everyone except those that match a pre-determined 'watch list' of known criminals.

Since September 11th, 2001, installations of surveillance-based systems in public places and airports have become hotly debated. On the one hand privacy advocates say that they are infringing on a person's rights and are not accurate enough to justify the cost; while on the other hand supporters say that if the system can lead to the capture of even one additional wanted terrorist it will be worth the investment. The debate continues, but an increasing number of airports are seriously considering installing such systems.

Work and environmental factors

Most commercially available systems currently require a number of factors to be taken into account when a user is being enrolled, verified or identified. The distance from the camera and the angle of the head need to be controlled to ensure that the captured image is of an optimum size. Some systems can identify a user at any angle as long as the camera can see both eyes while others need the person to be looking straight at the camera.

Ease of use

As most people are used to looking at a camera these systems are easy to use. All the person needs to do is stand in position, look at the camera and have their picture taken.

Performance

As with other biometric systems, facial recognition systems work best if a good image is captured at enrolment. For some, this means capturing a series of pictures of the same person with their head at a different rotation.

In surveillance applications, the performance varies widely depending upon sensible positioning of cameras. In this scenario, a high number of false alarms is clearly not desirable, and can be avoided if the tolerances of the system are increased – but of course this is at the risk of letting through people on the watch list. A sensible balance needs to be worked out, often depending on the resources available for checking out alarms when they occur.

6.1.6.5 Voice Based Verification Methods

Work has been ongoing in the field of speaker verification and recognition for almost 30 years. Texas Instruments first reported on its work in speech verification for access control as long ago as the early 1970s.

Speaker verification focuses on the vocal characteristics that produce speech and not on the sound or the pronunciation of the speech itself. This means that speaker verification systems cannot be compromised by mimics who reproduce the pronunciation of a word but cannot reproduce the vocal characteristics of the speaker. These vocal characteristics depend on the dimensions of the vocal tract, mouth, nasal cavities and the other speech processing mechanisms of the human body.

To prevent the password of an authorised user being digitally recorded and then played back to a speaker verification device, words may be picked randomly by the system from a selection stored at the time of enrolment.

Speaker verification systems come in two basic types: stand-alone units which can be networked within a building, and telephone-based systems. The economics for both types are quite different. For stand-alone systems, the cost is dependent on the number of access points that need to be protected while, for a telephone-based system, the cost of each access control point is negligible or even zero as often just a standard telephone handset is required. In a telephone system, however, the cost is in the central system software. Often this software will be operated by the supplier company and the customer will pay on a 'per transaction' basis.

Performance can also be radically different between the two types of system. Both are affected by the level of background noise, but there is an additional factor for telephone-based systems. Since the voice sample must be transmitted over telephone lines, performance can be affected additionally by the narrow bandwidth used for voice transmission over public telephone networks and, particularly, by any 'noise' on the line.

Beyond this, there is a further distinction with speaker verification systems between those that are text dependent, where only a pre-registered word, number or phrase can be used for verification, and text independent systems that can cope with a full vocabulary. Most speaker verification systems are text dependent.

The public acceptability and ease of use ratings of speaker verification systems are high. Background noise can be a major problem, however, leaving this biometric with a good false accept rate, but also with a high false reject rate. As with most biometric systems, a varying performance can be seen among commercial products.

Many companies have developed speaker verification as an 'add on' to their voice recognition technologies. In the past, a distinct line could be drawn to separate the companies in this industry into those that supply speaker verification over telephone lines and those that supply stand-alone systems. This is now changing as software solutions can be applied to either situation.

6.1.6.5.1 Operation of a Speaker Verification System

Text dependent verification enrolment on a text dependent system is carried out by speaking one or several words, or a series of digits, either of the user's or of the system's choosing, into a microphone. These usually need to be repeated a number of times so that an aggregate reading may be produced for a template.

Various acoustic characteristics are used to create the template. The pitch and the fundamental frequency, as well as formant frequencies, will be recorded along with energy readings and various other measurements. The characteristics of the word spoken are analysed and stored in the system, linked to a PIN or plastic card number for reference.

6.1.6.5.2 Requirements of a Speaker Verification System

Public Acceptability
The public acceptability of speaker verification is very high. Systems that rely on standard telephones have a slight edge but even those systems which perform speaker verification off-line with a special terminal are not generally intimidating. In a quiet office based environment, however, speaker verification systems may be slightly embarrassing to use.

Work and Environmental Factors

The problem of background noise can be overcome by some systems on the market. These rely on sampling the background prior to the user repeating the word. The background noise is then eliminated from the reading taken to give a clearer indication of the speech characteristics of the user. Most speaker verification systems update the speech templates each time a user is granted access. This is an important factor since it allows the system to cope with the onset of a cold by gradually amending the template as the cold worsens and then gradually restoring the template back to normal as the user recovers.

Ease of Use

These systems are particularly easy to use. Where more than one word has been stored as a template, the user will be informed of the word required either via a display or a pre-recorded voice. Problems with ease of use can arise with background noise either on the telephone line or at the user's location.

Circumvention

As previously mentioned, speaker verification is secure from the professional mimic since the systems makes a comparison of the word stored in a completely different way than humans compare voices. To prevent the use of a pre-recorded voice, words may be picked randomly from a selection stored at the time of enrolment, although this is clearly a potential vulnerability.

Performance

The problems of background noise mean that speaker verification systems have a reasonably low false acceptance rate, compared to their false rejection rates. As with all biometric systems, however, performance is improving.

6.1.6.6 Dynamic Signature Verification

With dynamic signature verification the comparison is not made between the way the new signature sample and the template look but between the way in which they have been written. The most obvious and important advantage of this is that a fraudster cannot glean any information on how to write the signature by simply looking at one that has been previously written.

The type of information which is gathered from the signature-writing process in order to carry out a verification varies between different suppliers' systems, but the data captured usually includes factors such as, the time taken to write the signature, the speed at which it is written, the number of times the pen is lifted from the paper and the points in time at which these lifts occur.

Approximately one per cent of the population vary their signature structurally by adding or omitting middle initials, dotted i's or underlines. In the past variations of this type have been rejected. However, some systems are now able to cope with these 'goats' by incorporating other measurements into their device. One such is to divide the signature into various segments, each being defined by the time between the pen being placed on the paper and lifted off again. Each segment can be individually compared to the template and extra or omitted sections can be ignored.

The majority of systems utilise a standard digitising tablet to capture the signature. These are sometimes found as standard computer peripherals. If someone is using a pen computer or a pen and tablet for computer entry, it is a reasonably easy task to install dynamic signature verification software. A few systems are available which use a special 'active' pen to capture the information.

6.1.6.6.1 Operation of a Dynamic Signature Verification System

To enrol onto a dynamic signature verification system, a number of signatures, on average about four, are recorded to form the template. Suppliers of signature verification systems take two approaches at this stage – either all of the signatures will be stored to show how the signature varies, as all do, or a composite of the signatures will be formed which shows the average signature of the individual.

For people whose signature varies greatly, the system may request that further signatures are given at the time of enrolment to produce a reliable picture indicating in which way the signature may vary.

Although there is no requirement for a visual signature, this is of importance for two reasons. Firstly, anyone who has tried to write with invisible ink will realise it is very difficult to do accurately. Secondly, with some applications, it is important that the user has a hard copy record of the operation that has been carried out for later reference.

6.1.6.6.2 Requirements of a Dynamic Signature Verification System

Public acceptability
The acceptability of signature verification in consumer applications is currently undergoing serious re-thinking. During the 1980s signature verification was seen by many as the most likely biometric to be used in consumer banking applications. This assumption is now being challenged as other biometrics, such as fingerprint verification, become more widespread. The recent signing of bills legalising electronic signatures, are expected to provide new momentum for the market.

Work and environmental factors
Many dynamic signature verification systems use standard computer peripherals to capture the signature details. Other companies have developed their own tablets on which a person must sign. Some of these require a stylus to be used when signing, while others allow a normal ballpoint pen to be used. The sensory equipment required can also be placed within a specially designed pen.

Ease of use
People are used to signing their name for many reasons, and most dynamic signature verification systems are no harder to use than a normal ballpoint pen. The 'active pen' type systems may take a short while to get used to signing with. Those with widely varying signatures may find the systems frustrating to use, as they will be repeatedly asked to re-sign their name.

Performance
The systems can be relatively difficult to break into, but the big variation in many people's signatures means that there is a relatively high FRR.

6.1.7 Biometric Applications

The uptake of biometrics is growing rapidly and the main areas of deployment are changing. The traditional stronghold use of biometrics in applications needing physical access control (and increasingly time and attendance monitoring) is now giving way to logical access control applications, where secure access to PCs, networks and intranets is becoming increasingly important. Transaction-based authentication is also providing momentum for the take up of biometric systems, especially in areas such as point-of-sale terminals, ATMs and e-commerce. Intelligent surveillance applications and citizen identification schemes are also becoming more prevalent in light of the terrorist attacks on September 11th.

The markets within which biometrics are used are diverse. The law and order market is steadily growing with applications, such as surveillance monitoring beginning to take off. One installation in the London Borough of Newham caused a notable reduction in crime. However, in the USA, systems in public places, such as at the US Super Bowl, caused a privacy backlash.

Prison visitor systems are also becoming more popular, where inmates, visitors and wardens have to use a biometric system to gain access to, or more importantly, to gain permission to leave the facility.

A fast growing application within the government sector is controlling immigration. Immigrant processing is one specific area that has seen an active involvement by the biometric industry recently. In the UK, for example, asylum seekers are now issued with a smart card containing their fingerprints in the card's chip. A database allows each new immigrant to be checked against previous records to discover whether this is a new immigrant to the system. New legislation surrounding immigration of normal passengers has also been introduced recently which could lead to an increased role for biometric technology in passport systems.

Within the travel industry, airports are particularly interested in using biometrics to secure and streamline passenger (and staff) flow from the airport entrance to the aeroplane and to enhance security.

The healthcare sector is providing many opportunities for biometric companies. In the USA, for example, legislation is requiring many healthcare providers to tighten up their data security arrangements and many are now turning to biometrics to do everything from securing individual PCs to securing of patient medical records. Hospitals are also seeing increased use of biometric solutions to increase physical security to secure areas.

One main area is the financial sector, where the use of biometrics includes cheque cashing, ATM money withdrawals, telephone banking and the securing of banks themselves. Another potential growth area in this sector is in point-of-sale terminals. Credit card manufacturers, such as Visa and MasterCard, have been particularly involved in investigating the potential of biometrically-secured payment mechanisms.

The potential for the use of biometrics in the telecoms sector is exciting, but has not yet emerged. One obvious application is the securing of mobile phones both to replace the PIN and to allow secure m-commerce. Integration issues and cost considerations are holding back the market so far, however. This area is one to watch in the near future as the price of biometric sensors and development of phones with video cameras (for face or iris recognition) increases.

The main biometric application areas are listed below:

- Financial
- Travel
- Government
- Healthcare
- Law and Order
- Education
- Leisure
- Telecom
- Commercial

A full market review is given in Chapters 2 and 3, describing the main drivers and revenue expectations in each of these sectors.

6.2 Glossary of Terms

Aberrant Data
Test data that is found to be anomalous and to have an irregular meaning in the context of that trial.

Access Control
The prevention of unauthorised use of a resource, including the prevention of use of a resource in an unauthorised manner.

Access Control List
A list of entities, together with their access rights, which are authorised to have access to a resource.

Accreditation
The procedure for accepting a system for use within a particular environment. Also the procedure of recognising both the technical competence and the impartiality of a test laboratory, which may be part of an independent organisation, to carry out its associated tasks.

Active Impostor Acceptance
When an impostor submits a simulated or reproduced Biometric Sample, intentionally attempting to relate it to another person who is an Enrolee, and

he/she is incorrectly recognised or Verified by a Biometric Verification Device as being that enrolee.

AFIS
Automated Fingerprint Identification System. The type of system used by law enforcement agencies to compare Fingerprints. Rather than performing a one-on-one match, these systems check a fingerprint against the whole database of stored prints and give a list of the most likely owners of the new print. Now also used in large-scale civil applications and to perform employee background checks as well.

Algorithm
A specified mathematical process for computation; a set of rules which, if followed, will give a prescribed result.

API
Application Program Interface. A 'layer' of program code which sits between a specific biometric device and an applications program, presenting a single sanitised interface to the application regardless of the type of biometric used.

Artificial Neural Networks
Computing method that emulates the way in which the human brain works; learning by past experience.

ATM
Automated Teller Machine. A machine which can handle many of the functions of a bank teller, including the dispensing of cash and often account balances and automated bill payments.

Audit Trail
A list stored by either the Biometric Verification Device or by another computer that gives details of who has tried to be verified at what time.

Authentic Acceptance
When a Biometric Transaction results in the correct recognition of identity, or the correct Verification of the claimed identity of an Enrolee.

Authentic Acceptance Rate
The probability (expressed as a percentage) that an Enrolee will be accepted as himself/herself by a Biometric Verification Device. (It is calculated as AAR = 100 – FRR, where AAR is the Authentic Acceptance Rate and FRR is the False Rejection Rate.)

Automatic Signature Checking
An automated means of checking the visual appearance of a signature, rather than the way in which it is written.

Authentication
The process whereby an individual, a card or a terminal is checked to ensure that he or it is a valid person or device.

Bar Code
A series of printed black and white lines of differing widths often found on cans in food stores. Each different width has a different meaning and when the whole code is scanned by a bar code reader, the complete information is read.

Beta Testing
Testing of a product which has not yet become commercially available. In terms of Biometric Verification Devices, these tests usually check the viability of the system and test for False Acceptance Rate and False Rejection Rate.

Biometric
A measurable, distinct physical characteristic or personal trait that can be used to recognise the identity, or verify the claimed identity, of an Enrolee.

Biometric Sample
A representation of a biometric that can be used as input to a biometric identification device.

Biometric Transaction
A sequence of Recognition attempts or Verification attempts culminating in the acceptance or rejection by a Biometric Verification Device of a person as an Enrolee.

Biometric Type
A wide variety of measurements are collected from and used by Biometric Verification Devices. These include Fingerprints, Finger Geometry, Palm Print, Ear Scanner, Retinal Scanner, Iris Scanner, Face Recognition, Dynamic Signature Verification, Blood Vessel Pattern, Voice, Spectral and Keystroke Dynamics.

Biometric Verification Device
An automated device capable of measuring the Biometric contained in a submitted Biometric sample, comparing the resulting measurement with that in one or more biometric reference Template, deciding whether they match sufficiently to the same person and indicating whether or not a Recognition or Verification of identity has been achieved.

Bit
Bit is an abbreviation of BInary digiT and is the fundamental building block of digital computer systems. A bit can be either a '1' or a '0' and several bits (usually in groups of eight) make up binary numbers (known as Bytes), which may represent an alphanumeric character, a value, a program instruction or other information.

Blood Vessel Pattern
The pattern of veins on the back of the hand which are used as a Biometric Type. The vein pattern is detected by an infrared scan of the back of the hand and classified according to the number, size and position of the blood vessels.

Booking
Law enforcement term for the process of documenting a suspect.

Byte
A byte is a group of Bits, usually eight.

Card Reader
A machine capable of reading and/or writing to a plastic card, such as a Magnetic Stripe Card or Smart Card.

Cashcard
A plastic card inserted into a Card Reader to obtain cash or services.

CCD
Charge-Coupled Device camera. A type of camera which takes digital pictures.

Central Database
Referred to when Templates are stored together on a PC or other computer, rather than in the memory of the Biometric Verification Device or on a Token.

Certification
The issue of a formal statement confirming the results of an evaluation, and that the evaluation criteria were correctly applied.

Claimant
A person who submits a Biometric Sample whilst claiming to be the legitimate Enrolee.

Claim of Identity
An Enrolee's submission of a Biometric Sample to a Biometric Verification Device for a verification decision.

CMOS
Image Sensing Technology CMOS stands for Complementary Metal Oxide Semiconductor and is one of several microchip fabrication technologies. In biometric usage, it allows for the integration of image sensing with image processing on the same chip and therefore has a lower cost than CCD systems.

Contactless Card
A card which can be read from a distance, rather than having to be inserted into or swiped through a Card Reader.

Data Broadcasting
A technique to enable information to be transmitted to special receivers on an unused portion of the TV waveband.

Decision Time
The precise time when a verification decision was made.

Demonstration Units
A mock-up of a Biometric Verification Device which is not yet commercially available. This device (usually more cumbersome than the end product) can be demonstrated to potential customers and other interested parties.

Detailed Design
A phase of the development process wherein the top level definition and design of the product/system under test is refined and expanded to a level of detail that can be used as the basis for implementation.

Development Process
The set of phases and tasks whereby the product/system under test is constructed, translating requirements into actual hardware and software.

Digitiser Pad/Tablet
A rectangular device that can pick up the measurements and image of a signature as it is written and delivers this information to a computer or processor in a form which it can understand. Some are specifically designed by developers of DSV systems, while others are standard computer peripherals.

DOS
Disk operating system.

DPI
Dots per inch. Term indicating image quality, referring to the number of Pixels in a digital image. The higher the number given for DPI, the more detailed is the resulting image.

DSV
Dynamic Signature Verification. Rather than comparing the visual image of the signature, dynamic signature verification systems compare how the signature is written; in other words the dynamics of the signature. These dynamics usually include the time taken to write the signature, positions at which the pen is taken off the paper, whether a stroke is done left to right or up and down etc.

Dwell Timing
The length of time in a Keystroke Dynamics system during which a key is depressed.

Dynamic Signature Verification
See DSV.

Ear Scanner
A Biometric Type.

Electronic Access Control System
This term includes Biometrics, Plastic Card and Pinpad secured entry systems – in fact any form of system which secures access electronically rather than just via a metal key and lock.

E-mail
Electronic Mail. Messages sent over a computer network, typically from user to user.

Electronic Nose
An electronic system which incorporates sensors to detect specific odours.

Encryption
A means of scrambling data so that it can only be understood by a party that has the key that changes it back to its original format.

Enhanced Magnetic Stripe Cards
The term enhanced magnetic stripe cards encompasses a range of products which incorporate some form of additional anti-counterfeiting technique, beyond that found on a standard bank card.

End-User
A person in contact with the product/system under test who makes use only of its operational capability.

Enrolee
A person who has a Template on file.

Enrolment
The process of collecting Biometric Samples from a person and their preparation and storage as a Template of that person's identity.

Enrolment Method
The specific method employed to collect Biometric Samples from a person during the Enrolment process.

Enrolment Time
The time period a person must spend to have his/her Template successfully created.

Equal Error Rate
The point at which the False Acceptance Rate and the False Rejection Rate are the same. This is usually quoted as a percentage.

Evaluation Unit
A device that can be bought to perform tests upon but that is not yet commercially available.

Exchange Authentication Information
Information exchanged between the claimant and the verifier during the process of Authentication.

Face Recognition
A Biometric Type that uses an image of the face or specific features of the face to verify a user.

False Acceptance
When a Biometric Transaction results in the acceptance of an impostor. (Also known as a Type II Error).

False Acceptance Rate
The probability (expressed as a percentage) that a Biometric Verification Device will fail to reject an Impostor. It is also known as a Type II Error Rate and is

calculated as FAR = NFA/NIRA (or NIVA) × 100 where FAR is the false acceptance rate, NFA is the number of False Acceptances, NIRA is the number of impostor recognition attempts and NIVA is the number of impostor verification attempts.

False Rejection
When a biometric transaction results in the failure to recognise the identity, or verify the claimed identity, of an Enrolee. (Also known as a Type I Error).

False Rejection Rate
The probability (expressed as a percentage) that a Biometric Verification Device will fail to recognise the identity, or verify the claimed identity, of an Enrolee. It is also known as a Type I Error Rate and is calculated as FRR = NRR/NERA (or NEVA) × 100 where FRR is the false rejection rate, NFR is the number of false rejections, NERA is the number of Enrolee recognition attempts and NEVA is the number of Enrolee Verification attempts

Feature Extraction
The process of obtaining particular details required from a larger image. Used particularly in face recognition.

Field Trial
Testing of a Biometric Verification Device using the general public or employees of a company in a situation which emulates a Live Application.

Finger Geometry
A Biometric Type whereby the geometry of either one or a number of fingers is used as the Biometric Sample. Some systems also incorporate the knuckles in the captured image.

Fingerprint
The pattern of ridges and valleys which cover the area from one side of the fingernail right around the pad of the finger to the other side of the fingernail. Within this area is found a wide variety of distinguishing marks from the major characteristics of arches, loops and whorls to the minor markings known as Minutiae.

Flight Timing
The time, measured by a Keystroke Dynamics system, between two keys being pressed.

Goats
Behavioural Biometric system users whose pattern of activity when interfacing with the system varies beyond the specified range allowed by the system, and who consequently may result in a False Rejection by the system.

Hacking
Illegal or unauthorised access to a computer or network from another computer via the telephone network.

Hand Geometry
A Biometric Type whereby the three-dimensional shape of the hand is used as the Biometric Sample. Just the geometry, showing lengths of fingers, shape and height of joints, is normally used rather than more detailed information.

Hands-free
Used to refer to the operation of a Biometric Verification Device which does not need the user to touch it (apart from placing the finger or hand on the scanning device). These have an auto-scan capability which can detect when a person is trying to use the system.

Hardware
Term referring to the actual 'box' and solid parts of a system rather than the programming Software which tells it how to function.

Henry System
The fingerprint classification system devised by Edward Henry of the Indian civil service in the 1860s.

Home Confinement
A system whereby people who have not committed too serious a crime can be restrained at home rather than being sent to prison. Speaker verification systems, for example, can be used to verify that the person is still at home via regularly placed telephone calls which the home detainee is required to answer.

Identification
When a new sample of a biometric is given and compared against a complete database of stored samples rather than compared to just one other Template. Identification systems, such as AFIS, provide a list of the most probable people that the given sample came from rather than Verifying that the new sample is from the person who it claims to be from.

Identity Badge
Any form of badge or tag which is in the possession of a user and indicates their ownership of it in some way. Identification may be in the form of a photograph or signature.

Image Processing
Term used to refer to any preparation that an image needs before it can be used as a Biometric Template.

Immigration
Application area including passport control and checking of travellers seeking asylum or trying illegally to enter a country.

Impostor
A person who submits a biometric sample in either an intentional or inadvertent attempt to pass him/herself off as another person who is an Enrolee.

Impostor Rejection
When a biometric transaction results in an Impostor not being accepted as an Enrolee.

Impostor Rejection Rate
The probability (expressed as a percentage) that an Impostor will not be accepted by a biometric identification device as an Enrolee. It is calculated as IRR = 100 – FAR where IRR is the Impostor rejection rate and FAR is the False Acceptance Rate.

Independent Organisation
An organisation that is separate and distinct from those responsible for the main development of the equipment under test, by ways of finance, management and other resources.

In-house Test
A test carried out entirely within the environs of the sponsoring organisation which may or may not involve external user participation.

Iris Scanning
A Biometric Type which uses the pattern on the iris of the eye as the means of comparison. The patterns compared include furrows, crypts, corona, filaments, pits, freckles, striations and rings.

ISO
International Standards Organisation.

KByte
Kilobyte. This is actually two to the power of 10 (i.e. 1024) Bytes – not simply 1,000 bytes as the term suggests.

Keystroke Dynamics
A Biometric Type which measures the way in which a person types. Each person has their own typing rhythm and it is this that is used to compare them with. Other measurements include Flight Timing and Dwell Timing.

Laboratory Test
A test, usually limited in scale, which is carried out entirely within a test laboratory, and in an artificially simulated test environment.

Latent Prints
Fingerprints found at the scene of a crime.

Live Application
A system which is up and running and being used on a permanent, fully functioning basis.

Logical Access Control
Access to computers and computer networks.

Liquid Crystal Display
A type of display found on most items nowadays including watches, calculators and Biometric Verification Devices.

Live Scan
Fingerprints taken via a scanner, from a live person. The alternative is to scan fingerprints from an image or a Tenprint Card.

Magnetic Stripe Card
The vast majority of these cards are financial transaction cards and, when this term is used without further explanation, an ISO standard financial transaction card is being described.

Matching Software
The part of a programme which performs the comparison between Templates and a new sample.

MB
Megabyte.

Minutiae
The small, detailed, identifiable features of a Fingerprint; terminations and bifurcation of ridges.

Minutiae Extraction Algorithms
A computing technique used to find and extricate the Minutiae data from a Fingerprint image.

MRTD
A Machine Readable Travel Document.

Mugshot
Law enforcement term for a facial image of a person.

Network
Infrastructure connecting Biometric Verification Devices or computers together.

Network Security
General term used when referring to the securing of access to a Network.

OCR
Optical Character Recognition. Machine readable characters.

OEM
A company which buys components from a number of other companies and integrates them together into a single product.

Off-line
In the card arena, the term off-line refers only to the authorising of a transaction or an entry to a building and requires that such operations are carried out without referral to any other part of the Network.

On-line
This refers to any system where the individual components are connected via telecommunications lines either directly to each other or indirectly via a switching centre. In the biometrics arena, it is used to refer to a system where the Biometric Template is transmitted through the telecommunications network to the Biometric Verification Device at a different location.

Optical Assembly
The part of certain Biometric Verification Devices that captures the image from the hand, finger, etc. One system utilised by Fingerprint verification devices involves a prism, against which the finger is placed, and a light beam.

Optical Memory Card
These cards have very high memory capacities in comparison to other cards of ISO dimensions. In order to write information to the card, a laser beam is used to selectively burn pits into an area of reflective material located on one side of the card. When the card is inserted into a read/write unit, a laser beam detects the presence or absence of a pit (by reflection) and interprets this as representing either a '1' or a '0'. This same laser is also used to write information on the card once it is in circulation.

OS/2
A computer operating system.

Palm Print
A Biometric Type that compares the pattern made by the creases in the palm of the hand.

Passive Impostor Rate
When an Impostor submits his/her own natural Biometric Sample, either intentionally or inadvertently attempting to relate it to another person who is an Enrolee, and he/she is incorrectly recognised or verified by a Biometric Verification Device as being that Enrolee.

PCMCIA Card (or PC Card)
A class of peripheral devices named after the Personal Computer Memory Card International Association, a body originally set up to define a standard for the use of 68-pin memory cards in personal computers. The cards are easily slotted into a computer in order to enhance either its memory or performance, or to enable it to perform a special function. The cards are the shape of a credit card but are considerably thicker and have a 68 pin socket along one end.

Penetration Testing
Tests carried out by an evaluator on the product/system under test in order to confirm whether or not known vulnerabilities are actually exploitable in practice.

Performance
The effectiveness and efficiency of the system under test when measured against the criteria established prior to the commencement of the test (and which by agreement of all parties involved may have been subsequently varied).

Performance Criteria
Pre-determined criteria established to evaluate the performance of the system under test.

Personal Authenticating Information
Information used to Authenticate a user's identity. The information can be derived from something the user knows (such as a secret password), something the user has (such as exclusive possession of a badge), something the user is (such as a Fingerprint), or any combination of the three.

Pilot Project
A test of a Biometric Verification Device as part of a complete system. These tests are usually performed to evaluate the various parts of the system and allow changes to be made before it becomes a Live Application.

PIN
Personal Identification Number. Usually a four-digit number which is issued to a person to connect them to their Template. It is also used in banking applications to Authenticate a bank card at an ATM or at the point of sale.

PINpad
Usually a small device incorporating keys for the numbers '0' to '9' and used for the entry of a user's PIN.

Pixels
The finest detail that can be recorded on the recording medium. One dot on a computer screen.

Portable Template Carrier
A Token, usually in the form of a Plastic Card, which can be carried by the owner. On its memory is stored the owner's Template.

Principal Component Analysis
A method used in face recognition to manipulate and compare two faces.

Production Unit
A Biometric Verification Device which is commercially available.

Prototype
A Biometric Verification Device which is still being developed. These units are designed to evaluate an idea and the work done so far.

Public Test
An open test carried out within a public environment where the conditions prevailing are not totally under the control of the sponsoring organisation or the test agency.

Recognition
see Identification.

Remote Banking
A method of performing banking transactions from another location. This can be done through a computer, over the telephone or by mail.

Remote Storage
When Templates are stored together on a computer at a different location to the Biometric Verification Device, rather than in its memory or on a Token.

Response Time
The time period required by a Biometric Verification Device to complete a Biometric Transaction.

Retinal Scanner
A Biometric Type which compares the pattern of blood vessels at the back of the eye. A scan is performed, centred on the area of sharpest sight, the fovea.

RF-ID Card/Tag
Radio Frequency-IDentification. RF-ID is a commonly used means of allowing tags and cards to communicate over distances of more than a few centimetres.

Ridge Flow Pattern
The major pattern on the Fingerprint caused by the ridges 'flowing' in the same direction.

Robustness
The inherent strength of a system which enables it to cope with unspecified conditions or events without compromise.

Rolled Fingerprint
A Fingerprint image derived by the finger being rolled sideways from one nail edge to the other to give an image of the whole fingerprint.

Screening Test
A test of just a biometric device alone, as opposed to its use within a complete integrated system.

Self Proving
The ability of a system to perform pre-determined tests that enable it to assess its own integrity and report thereon; such testing may be triggered automatically or as a result of direction from the system operator.

Self Testing
An internally directed test carried out by the system operator in order to evaluate the continuing integrity of the system in use.

Smart Card
Credit card sized plastic cards in which a chip has been embedded. The chip is usually embedded towards the middle of the left hand side and may be capable of performing data processing. This chip usually communicates with the outside world via metal contacts on the face of the card. These cards are more secure

than Magnetic Stripe Cards, have a higher storage capacity and are often capable of processing data.

Software
The part of a system which is not the Hardware. The part which performs the processing and tells the system exactly how to perform and when.

Speaker Identification
The technology of comparing the vocal characteristics of an unknown speaker against a database of known speakers.

Speaker Verification
A Biometric Type which compares the vocal characteristics of the user. Rather than comparing the words spoken, these systems compare the way in which they are said. The voice is made up from a number of characteristics which include the size and shape of the mouth, vocal tract and nasal cavities.

Spectral Characteristics
A Biometric Type which recognises a person by measuring the composition and structure of their skin and underlying tissue.

Speech Pattern
The wave pattern that is produced from the timbre of the spoken phrase.

Speech Processing
Term used to refer to any mechanical or electronic system which processes speech. These include call processing and automatic fax sending.

Speech Recognition
Systems which can recognise either specific spoken words and phrases or a whole vocabulary and react to what is being said.

Sponsor
The person or organisation that requests that a test is carried out.

Stand-alone
A system that works completely on its own without needing to be connected to another machine.

System Configuration
The arrangement of, and the relationship between, the various components of the system.

Systems Integrator
A company which installs complete systems, usually using parts from a number of suppliers.

Tamper Resistance
The physical capability of components within a system to withstand external attack and, if necessary, to destroy any confidential information contained therein.

Telephone-based Speaker Verification
A system whereby the voice is Verified by a computer located at a different place from the speaker. The two are connected by the telecommunications infrastructure.

Template
A data set representing the Biometric measurement of an Enrolee which is maintained on file and used by a Biometric Verification Device for comparison against subsequently submitted Biometric Samples.

Template Dormant Time
The elapsed time between the creation, or last update, of a Template and its current use.

Template Identifier
The item used to tie a user to their own template to make sure the correct template is called up when required; this can be a PIN, password or card number.

Template Matching
Where the newly captured sample goes through the same algorithms as the sample captured during the Enrolment process.

Template Maturity
The number of Biometric Samples, including the original sample, contributing to the Template currently on file.

Tenprint Cards
Paper documents used in law enforcement to keep a record of all ten fingerprints of an individual.

Test Regime
The rules which govern the organisation of a test and how it should be carried out.

Text Dependent
A Speaker Verification system that requires the user to speak only a pre-registered word, number or phrase.

Text Independent
A Speaker Verification system that can cope with a full vocabulary.

Threshold
A moveable setting on a Biometric Verification Device which enables the system to allow in either more or less Impostors. What has to be remembered is that, as the False Acceptance Rate improves, the False Rejection Rate decreases – and vice versa.

Throughput
The number of Biometric Transactions that a Biometric Verification Device completes within a stated time interval.

Tied Distributor
A company that has signed an agreement to sell the products of another company.

Time & Attendance
The application whereby a Biometric Verification Device is used to Verify a user before the system will accept that they have arrived at work. Since the person also has to be Verified when they leave, the employer can be certain that the person was working that day and knows how many hours they worked and therefore how much money to pay them. The use of Biometrics ensures that no-one else can punch in for them.

Timeliness
A method to prevent a valid Authentication from being replayed at a later time by using a probe of information as a challenge, requesting a proper and timely response.

Token
A possession which shows the identity of its owner, such as an Identity Badge.

Tolerance
A degree of tolerance is incorporated into biometric systems to allow for a less than 100% match to be Verified.

Tone Pad
Usually a small device incorporating keys for the numbers '0' to '9'. The key for each number produces a different note which can be recognised by other systems. Currently used in some Remote Banking applications to transmit the account and pass number down the telephone line to the bank's computer.

Total Internal Reflection Spectroscopy
A technique used in fingerprinting to distinguish between a live human finger and an imitation, such as pig skin or latex.

Turnkey System
A complete system that has been developed by a company for a specific application area. The off-the-shelf system can then be purchased by anyone for their own use.

Type I Error
See False Rejection.

Type I Error Rate
See False Rejection Rate.

Type II Error
See False Acceptance.

Type II Error Rate
See False Acceptance Rate.

Uniformity of Errors
Consistency of errors reported within a test which facilitates their resolution.

Uniformity of Performance
An objective of the system operator whereby random occurrences and unknown problems do not arise.

Unix
A computer operating system.

USB
A Universal Serial Bus replaces all the different kinds of serial and parallel port connectors with one standardised plug and port combination. It allows PCs and peripherals to be connected in an easy way.

Validation
The process of demonstrating that the system under consideration meets in all respects the specification of that system.

Verification
The process of comparing a submitted Biometric Sample against the Template of the single Enrolee whose identity is being claimed in order to determine whether it matches the Enrolee's template or not.

Verification Attempt
The submission of a Biometric Sample, accompanied by a claimed Enrolee identity, to a Biometric Verification Device for a Verification decision.

Verifier
An entity which is, or represents, the entity requiring an Authenticated identity. A verifier includes the functions necessary for engaging in Authentication exchanges.

Voice Mailbox
A place where verbal messages can be left for a specific person to collect at a later date.

Volatiles
The chemicals making up a person's smell which are emitted from pores all over the body.

Vulnerability/ies
Security weakness/es in the product/system under test (such as failures in analysis, design, implementation or operation).

Welfare
The application area whereby a new claimant's Biometric Sample is compared against the whole database of samples from other claimants to ensure that the new person is not already claiming benefit under a different name or from a different office.

Windows
A computer operating system.

Zero Effort Forgery
An arbitrary attack on a specific Enrolee identity where the Impostor masquerades as the claimed Enrolee.

6.3 Standardisation Update

6.3.1 Introduction

For many years the development of a unified Biometric Application Programming Interface, and the respective biometric template formats, was considered vital if there was to be widespread adoption of biometric technology. So, two years after the first release of BioAPI and CBEFF (Common Biometric Exchange File Format), is the work almost finished?

The short answer to this question is "no". But there have been significant developments, and most noticeably there has been a shift in focus from a technology-orientated approach to an application and security integrated approach. Ongoing work is now aimed at accelerating the integration of existing standards (rather than developing new ones) into products and solutions, as well as improving interoperability between application domains.

This part of the appendix was formulated with the kind help of Infineon Technologies' Dr Brigitte Wirtz, who is an active member of the BioAPI and CBEFF consortia. It gives an overview of the major biometric standardisation initiatives, including a view on interoperability, and focuses on their most recent advances and achievements.

6.3.2 Biometric APIs

Generally speaking, a biometric Application Programming Interface (API) separates individual biometric technologies from applications. In other words it allows the respective technologies to be used across multiple applications as well as the simple integration of that technology into existing applications.

Together with standardised biometric data formats, the biometric API is the basis for 'plug & play' integration of biometrics into applications and security concepts, especially in the internet world. Plug and play is essential for widespread adoption by integrators. Integrators should not need specialist knowledge in order to integrate biometrics into their applications. Integrators should also be able to easily replace different sensors or software versions (from varying suppliers), or combine a number of biometrics in a standardised form.

6.3.2.1 BioAPI

The platform-neutral BioAPI standard governs the introduction of biometrics into standalone or client-server applications. It evolved from a number of competing biometric API proposals including: *SVAPI* (Speaker Verification API); *BAPI* (I/O Software's Biometric API); *CAPI* (IBM's Advanced Identification API); *BAAPI* (a commercial biometric API); and *HA-API* (User Authentication API).

BioAPI provides a high-level generic biometric authentication model, suited for any form of biometric technology. It covers the basic functions of enrolment, verification and identification and includes a database interface to allow a biometric service provider (BSP) to manage populations of identities.

With these convenience functions, application programmers get an interface that allows them to concentrate on their application instead of the internals of biometrics. Additionally, however, BioAPI also provides primitive functions which allow for more flexible system arrangements – for example by enabling the capture of samples on a client, while enrolment, verification and identification take place on a server.

In the history of biometric standardisation, the BioAPI has been an unparalleled industry effort for the open and interoperable standardisation of the biometric API. A primary aim has been to ensure interoperability with all other relevant standardisation initiatives, such as CBEFF, ANSI X9 and ISO smart card standardisation.

At the end of March 2001, the BioAPI consortium released the updated *BioAPI V1.1*, which included its *Windows* platform reference implementation. The specification was released as an ANSI/INCITS standard in February 2002. Because the specification was initially designed as platform independent, the consortium is currently working on *Unix* and *Linux* reference implementations. For a brief overview of the history and status of BioAPI see Table 6.1.

Table 6.1 Brief History and Status of the BioAPI

April 1998:	founded by Compaq, original promoters Compaq, IBM, Identicator, Microsoft, Miros, Novell
Dec. 1998:	merger of BAPI and Bio-API
March 1999:	merger of HA-API and Bio-API
	Membership continues to grow (currently over 90)
30 March 2000:	version 1.0 of specification released
September 2000:	reference implementation completed
March 2001:	specification and reference implementation version 1.1 released
February 2002:	accepted as ANSI/INCITS 358-2002
	INCITS M1 committee letter ballot resolution to send BioAPI through ISO fast track process once ANSI publishes
	Products being announced
Ongoing projects:	Conformance Test Suite, Platform portations

Today, support for the BioAPI standard is significant with more than 90 member companies and organisations involved. Some end users have given strong backing to the standard. The US Department of Defense, for example, requires BioAPI compliance as a pre-requisite before it will consider a biometric solution.

6.3.2.2 Java Card Biometric API

BioAPI and CBEFF were originally designed to help integrate biometrics into client-server scenarios – in particular to introduce biometric identification into internet applications. More recently that position has shifted to a more solution and application oriented approach, where biometrics are just a small portion of the overall solution, and more diverse system configurations are considered.

One such configuration is the storage of biometric data in a smart card or token that is itself integrated into an overall client-server scenario.

The Java Card Forum's Biometric Taskforce has decided to define a more generic Biometric API to enable the open platform and interoperable integration of biometrics into smart cards. (In contrast, ISO/IEC Subcommittee 17 (SC17) – the body responsible for the international standardisation of personal identification and cards – has chosen to use and adapt existing card commands for biometrics.)

Working in close co-operation with the respective Biometric Consortium Working Groups, the primary goal of the Taskforce is to describe the rationale and architecture of a Java Card biometric API.

While the original members are the classic smart card manufacturers Schlumberger, Oberthur, Hitachi and GemPlus, its target is to define a biometric card API, that:

- uses Java Card and its security mechanisms for storage of all sensitive information on the card;
- enables efficient matching-on-card;
- allows for multiple templates and biometrics;
- enables different applet implementation scenarios from standalone to server and client applets.

Integrating this work with the existing CBEFF and BioAPI standards will ensure the transparent integration of a card or token platform into a networked world, and thus enables seamless implementation of biometric solutions over a set of distinctive computing platforms.

A common White Paper on the JavaCard Biometric API jointly written by the Biometric Consortium Interoperability, Assurance, and Performance Working Group and the Java Card Forum's Biometric Taskforce is currently under review and was supposed to be submitted as an official working group document by September 2002.

6.3.3 Biometric Data Interoperability

The interoperability of biometric data is one of the biggest issues for biometric standardisation. As data collection is an expensive task, it could be desirable for the same enrolment data to be used by different biometric identification products. (Given the appropriate data and privacy protection measures are put in place.)

Biometric data sharing between different system components in a compound system – for example the seamless integration of biometrics into smart card integrated applications – requires interoperability of data.

Interoperability can be achieved in two ways – one being the dedicated standardisation of the biometric template data, the other being the standardisation of the representation form of the data only. While the first alternative is an approach that has been undertaken for some dedicated applications, such as the use of minutiae-based fingerprint data for AFIS identification or driver's licenses, the second approach is technology neutral.

This appendix only looks at technology neutral biometric template standardisation. There is not yet a single industry-wide biometric data standard, and again some of the recent standardisation work stems from the desire to integrate biometric identification into broader system scenarios, such as making the smart card available as a secure biometric data storage and computing device, as well as using biometrics in the context of traditional internet communication schemes.

Three major standards are already available that achieve interoperability when describing biometric data – the Common Biometric Exchange File Format (CBEFF); the BioAPI biometric data format (BIR); and the ANSI X9.84 Biometrics Management and Security for the Financial Services Industry data format.

6.3.3.1 The BioAPI BIR

The BioAPI biometric data format is designed as a fixed format biometric identification record and is known as the *BIR* (Biometric Identification Record). It consists of a header, the opaque biometric data and an optional signature.

The 16 byte header holds information, such as the length of the complete BIR; the header version number; the type of data; the format ID of the opaque data format (which can either be proprietary or in an open format); the quality and purpose of the data; the biometric modalities (finger, face, voice, etc.); and the specifications of an optional signature.

6.3.3.2 CBEFF

The Common Biometric Exchange File Format was predominantly defined by a subgroup of BioAPI consortium members – originally Identicator, Infineon Technologies, NIST/ITL, NSA, Saflink and Veridicom. More recently CBEFF work has been a joint effort of the industry, user organisations and the Biometric Consortium in co-ordination with the BioAPI consortium, ANSI X9F4 Financial Services Working Group, the IBIA and the interfaces group of the German TeleTrusT.

Chaired by NIST/ITL the subgroup generated a 'technology-blind' standard file format to facilitate exchange and interoperability of biometric data. Because it is a data description format, it does not standardise the underlying biometric data sets – therefore it is not a barrier to competition between the providers.

The data structure within the CBEFF specification was designed to handle a wide variety of application systems, from sophisticated systems with complex biometric procedures (that might require a complex structure as well as compliance with an internationally accepted data encoding scheme), to systems that have limited storage media and require a data structure with minimum overhead. Template description formats in CBEFF – so-called Patron formats – always describe the specific encoding of the data elements, additional (non-common) data elements, as well as possibly optional fields.

The concept is more general than the original BioAPI BIR definition. In fact the CBEFF serves as the biometric template definition for BioAPI.

In a similar manner to BioAPI's BIR, the CBEFF data format comprises a header followed by an opaque data format.

Currently two biometric Patron formats are defined in the CBEFF. One of these is a 'simple' BioAPI-like format which uses fixed length fields of one or two bytes in length. The other is the ASN.1 (Abstract Syntax Notation One) Patron, which is the definition of a biometric data set in the standardised description language, which is also used by X9.84.

Both equivalent encodings are given in the CBEFF document, which was released as a NIST/ITL standard in November 1999.

The recent demand for application integration with respect to biometric data interoperability has led to the definition of two new Patron specifications: a smart card format using TLV encoding present in the smart card world, as well as an XCBF patron, describing a non-binary patron format to be used in XML systems and applications.

6.3.3.3 XCBF

Since spring 2002 OASIS, ISO/IEC and the Biometric Consortium have cooperated to define XCBF, a biometric XML schema to govern the exchange of biometric data in e-commerce. XML is fast becoming the standard language of the Internet and so XCBF is designed to overcome the shortfalls of the existing binary-based CBEFF specifications in these sort of applications. Interoperability with existing data formats and security standards will be guaranteed.

If all goes to plan a draft version of a working module will be ready by November 2002, the draft version of example binary encodings of XML values by March 2003, the draft final versions of all remaining deliverables by May 2003, the final versions of any new CBEFF formats by July 2003, and the final versions of all deliverables by November 2003.

6.3.3.4 X9.84

The X9.84 (Biometric Information Management in Financial Services) definition for biometric data uses ASN.1 encoding rules. These rules are popular in the smart card sector and in security information processing standards such as X.509, SET and Microsoft Active Directory. The advantage of ASN.1 encoding of data formats lies in its platform independence and its flexibility with respect to the size of individual data fields. It is also flexible when introducing new data fields.

As one of the CBEFF patrons, the X9.84 biometric data format is naturally CBEFF and BioAPI compliant. Furthermore, biometric data is stored in the same format as X.509 version 3 certificate extensions, and as such is compliant with the Internet Engineering Task Force's (IETF) PKIX-standards (Public Key Infrastructure for X.509) for the integration of biometric and cryptographic keys in certificates.

6.3.3.5 ISO/IEC 7816-11

New ISO/IEC international IT standardisation activities are usually decided without major controversial discussion. For biometrics this has been surprisingly different. Subcommittee 17 (SC17) of ISO/IEC Joint Committee 1 (JTC1), which is responsible for the international standardisation of personal identification and cards issues, has developed standards with biometric-related elements for three years. Meanwhile, Subcommittee 27 (SC27) is responsible for biometrics standardisation related to security issues.

Soon after the tragic events of September 11th, the USA felt that the biometric standardisation of identification documents and cards should have its own Working Group (WG) within SC17. The USA applied for such a WG at SC17's Plenary in October 2001 and gained unanimous agreement – hence the formation of WG11.

At the end of 2001, however, the USA changed its position on WG11 and instead sought the establishment of a new Subcommittee within JTC1. A ballot to create this Subcommittee (SC37) was held in early June this year and was passed with 11 approvals and 9 disapprovals. The subcommittee's inaugural meeting will probably be held in December 2002.

SC37 will aim to avoid standardisation issues being considered by other sub-committees, in particular SC17 and SC27. So, when neither identification documents and cards nor security techniques are involved, SC37 will concentrate on issues related to application program interfaces, file formats, biometric template standardisation and application/implementation profiles.

The new biometrics Working Group WG11 of SC17 (Secretariat and Convener provided by Germany) started its activities at the end of June 2002. Its first standards development work is concentrated on logical data structures with relevance to biometrics features for identification documents. This task is supported by a few governments and by the International Civil Aviation Organisation (ICAO).

The end result is that biometric standardisation will be undertaken in different subcommittees, each with their own agendas. Close and constructive cooperation will no doubt be the best way to ensure biometric related standards are progressed quickly.

Aside from this confusion over new activities, significant progress has been achieved on the "Personal verification through biometric methods" Work Item – ISO/IEC 7816-11. It is now a Final Committee Draft and is currently under vote. ISO/IEC 7816-11 defines commands and data elements for biometric user verification on a smart card. The idea is to use existing smart card structures and upgrade the given smart card commands to allow for the introduction of biometrics as a personal identification method.

The biometric data structures are currently being aligned with the CBEFF and X9.84, with the ultimate aim of becoming a CBEFF patron in the future. The interoperability or compliance of ISO/IEC 7816-11 with the ICAO logical data format is still a topic that has to be looked at and will certainly be targeted by WG 11. Another New Work Item Proposal from Germany's DIN relating to the matching of biometric data on smart cards is also being considered.

6.3.4 Application Integration

The integration of a biometric API into the target platform's operating system is a necessary step for the security integration and ubiquity of biometric systems. The security of the biometric data, if stored on the platform's file system, as well as the security mechanisms between the different biometric functions and modules cannot be viewed independently from the security policies and facilities of the host operating system. Biometrics must be integrated into the complete software stack of the respective platform API to achieve integrated security.

A biometric API only offers a unique interface to the application. Biometric system integration – from BIOS integration up to data protection and integrity checking within the operating system – is crucial to ensure overall security, and should not be left to the application or to a proprietary implementation within the biometric service provider (BSP).

Rather than implementing security within each biometric API, the biometric API interface should be able to combine securely with accredited security tools, of which there are several on the market. It would make sense, for example, that a BSP can interface to existing cryptography toolsets, such as CAPI, PKCS or GSSAPI, rather than each BSP implementing cryptography on their own.

6.3.5 CDSA

The Open Group CDSA (Common Data Security Architecture) is a standard, that implements user authentication within a security framework via a layered approach that bridges from special *Security Add-In Modules* via a *Common Security Services Manager* to the *System Security Services*, that are finally offered to the applications. In CDSA user authentication is handled via UAS (User Authentication Services). Thus biometrics can be integrated into an overall security architecture in a consistent and secure manner – a BSP just plugs into the UAS in the same way as classical user authentication mechanisms do.

6.3.6 Microsoft

The announcement by Microsoft in May 2000 that it would foster the wide-spread growth of biometrics through the integration of biometric technology in future versions of the Microsoft Windows operating system was a promising sign for the push of biometric technology. But, a major part of the industry wondered why Microsoft did not choose to integrate BioAPI, the industry-wide acknowledged standard for a biometric API. Instead it decided to co-operate with I/O Software, licensing its commercial (and not BioAPI compliant) BAPI.

Now that the dust has settled, Microsoft's choice of API is not necessarily as big a rift in the world of standardisation as it may at first have seemed, especially considering that the interoperability of data might be even more important than the selection of API, which could at least be solved by an appropriate translation layer between BioAPI and the Microsoft BAPI.

Non-NDA meetings between the BioAPI consortium and Microsoft have not progressed too much. Whereas Microsoft XP still does not hold any biometric features, public rumours now say that the next Service Pack or LongHorn are going to have biometric interfaces included.

Microsoft is believed to have completely reworked BAPI. The MS architecture probably extends the specification of a top level biometric API by adding a lower level specification for fingerprint devices (image, self-encoding and self-matching), a ResourceManager for the maintenance of biometric devices, a device class 'Biometrics' as well as a WHQL specification for such devices. Biometric service providers could thus be loaded above the ResourceManager.

Given that the standard can quickly be made interoperable with the existing BioAPI and CBEFF standards, this architectural integration into the pre-dominant MS platform could be a further significant step towards the ubiquitous introduction of biometrics as a means of person identification on computer platforms.

6.3.7 X9.84

The scope of X9.84 is the use of biometric technology for identification and authentication of banking customers and employees. But far from just providing a CBEFF/BioAPI compliant definition of the biometric template data, the ANSI standard serves as a security integration profile for biometric application integration. As such it deals with the security and management of biometric data and physical hardware, the application of biometric technology for logical and physical access, the encapsulation of biometric data as well as with techniques for secure transmission and storage of biometric data.

The target was to establish minimum security requirements for the effective management of biometric data, such as, to maintain the data integrity of biometric data and verification results between any two components; to authenticate the source of the biometric data and verification results between the sender and receiver component, as well as to ensure the confidentiality of the biometric data during transmission.

The X9.84 could probably be called the first available biometric security and application integration profile.

6.3.8 Missing Links

Despite excellent progress in the last two years, standardisation activities are still in flux. This is partially due to a natural transition from technology standardisation to an application and solution-oriented approach. The standards must now interface with interfaces already existing in the end-user's world. This is why, for example, the INCITS Technical Committee Biometrics M1, that has recently been established as the generic biometric working group within ANSI, is not only striving for fast adoption of existing standards like BioAPI and CBEFF through fast-track ISO procedures, but is also looking at application profiles and implementation agreements.

Besides the general application shift, that has occurred in the biometric industry, there are further issues, such as biometric evaluation, certification and assurance and privacy and data protection that need serious consideration. It is likely that these will demand standardisation, as well, before the picture is complete.

Biometric performance and security is one major criteria governing an application integrator's choice of a biometric solution. Although a lot of work has been done by well-known institutions in the past, an internationally agreed and standardised procedure for testing biometric systems does not exist. The formulation of Biometric Protection Profiles according to Common Criteria is under way – and hopefully the three versions that are currently under review – from Canada, the UK and the US – will finally converge.

6.3.9 Summary

The considerable amount of standardisation work mentioned above might create the impression that the industry has only just begun on the road to standardisation. However, the message is that recent developments are a sign of convergence, with a striving for interoperability on existing standards and a focus shift toward integrated solutions.

Major standardisation efforts are aligned with each other with respect to templates, API and security architectures. Different biometric standards, such as the BioAPI, ISO/IEC 7816-11, PC/SC and the JavaCard Biometric API, for example, fit into the same consistent "big" picture of a distributed secure systems application with biometrics.

The transition of industry consortia standards to internationally accredited standards has started and is continuing. Any standardisation work will never totally end as technology always undergoes a natural improvement cycle. But biometric standardisation has definitively entered a phase of consolidated interoperability and gradual improvement.